GREEN PRODUCTION ENGINEERING AND MANAGEMENT

Woodhead Publishing Reviews: Mechanical Engineering Series

GREEN PRODUCTION ENGINEERING AND MANAGEMENT

Edited by

CAROLINA MACHADO

School of Economics and Management, University of Minho, Braga, Portugal

J. PAULO DAVIM

Department of Mechanical Engineering, University of Aveiro, Aveiro, Portugal

ELSEVIER

WP WOODHEAD PUBLISHING

An imprint of Elsevier

Woodhead Publishing is an imprint of Elsevier
The Officers' Mess Business Centre, Royston Road, Duxford, CB22 4QH, United Kingdom
50 Hampshire Street, 5th Floor, Cambridge, MA 02139, United States
The Boulevard, Langford Lane, Kidlington, OX5 1GB, United Kingdom

Notices
Knowledge and best practice in this field are constantly changing. As new research and experience broaden our understanding, changes in research methods, professional practices, or medical treatment may become necessary.

Practitioners and researchers must always rely on their own experience and knowledge in evaluating and using any information, methods, compounds, or experiments described herein. In using such information or methods they should be mindful of their own safety and the safety of others, including parties for whom they have a professional responsibility.

To the fullest extent of the law, neither the Publisher nor the authors, contributors, or editors, assume any liability for any injury and/or damage to persons or property as a matter of products liability, negligence or otherwise, or from any use or operation of any methods, products, instructions, or ideas contained in the material herein.

Library of Congress Cataloging-in-Publication Data
A catalog record for this book is available from the Library of Congress

British Library Cataloguing-in-Publication Data
A catalogue record for this book is available from the British Library

ISBN: 978-0-12-821238-7 (print)
ISBN: 978-0-12-821239-4 (online)

Publisher: Matthew Deans
Acquisitions Editor: Brian Guerin
Editorial Project Manager: Chiara Giglio
Production Project Manager: Anitha Sivaraj
Cover Designer: Christian J. Bilbow

Typeset by STRAIVE, India

Contents

Contributors

Mohammed El Amine Abdelli
University of Western Brittany-UBO, Brest, France

Rohit Agrawal
Department of Production Engineering, National Institute of Technology, Tiruchirappalli, Tamil Nadu, India

Soudani Ahlem
University of Guelma, Guelma, Algeria

Sudin Bag
Department of Business Administration, Vidyasagar University, Midnapore, West Bengal, India

Arup Barman
Department of Business Administration, Assam University, Silchar, Assam, India

Sauro Bianchi
Bern University of Applied Sciences, Department of Architecture, Wood and Civil Engineering, Biel, Switzerland

Nedjari Fatima
Djillali Liabes University, Sidi Bel-Abbes, Algeria

Diana Fernandes
University of Minho, Braga, Portugal

Jan Frecè
Bern University of Applied Sciences, Business School, Bern, Switzerland

Deane Harder
Bern University of Applied Sciences, Business School, Bern, Switzerland

Marilyn M. Helms
C. Lamar and Ann Wright School of Business, Dalton State College, Dalton, GA, United States

Aref Aghaei Hervani
Lander University, Greenwood, SC, United States

Michail Kyriazopoulos
Bern University of Applied Sciences, Department of Architecture, Wood and Civil Engineering, Biel, Switzerland

Carolina Machado
School of Economics and Management, University of Minho, Braga, Portugal

Aied Malika
University of Lounici Ali de Blida 2, El Affroun, Algeria

Santosh Nandi
Walker College of Business, Appalachian State University, Boone, NC, United States

Nilanjan Ray
Marketing and Examination Coordinator, Adamas University, Kolkata, West Bengal, India

Biswajit Roy
Future Business School, Kolkata, West Bengal, India

Joseph Sarkis
School of Business, Worcester Polytechnic University, Worcester, MA, United States

S. Vinodh
Department of Production Engineering, National Institute of Technology, Tiruchirappalli, Tamil Nadu, India

Arabeche Zina
Center University Ahmed Zabana Relizane, Relizane, Algeria

Editors' biographies

Carolina Machado

Carolina Machado received her PhD degree in Management Sciences (Organizational and Policies Management/Human Resources Management) from the University of Minho in 1999, master's degree in management (Strategic Human Resource Management) from the Technical University of Lisbon in 1994, and a degree in Business Administration from the University of Minho in 1989. She has been teaching human resources management subjects since 1989 at the University of Minho and has been an Associate Professor since 2004, with experience and research interest in the fields of human resource management, international human resource management, human resource management in SMEs, training and development, emotional intelligence, management change, knowledge management, and management/HRM in the digital age. She is Head of the Human Resources Management Work Group at the School of Economics and Management at University of Minho, Coordinator of Advanced Training Courses at the Interdisciplinary Centre of Social Sciences, Member of the Interdisciplinary Centre of Social Sciences (CICS.NOVA.UMinho), University of Minho, as well as Chief Editor of the *International Journal of Applied Management Sciences and Engineering* (IJAMSE), guest editor for journals, books editor and book series editor, as well as reviewer for prestigious international journals. In addition, as editor/co-editor or as author/co-author, she has also published several books, book chapters, and articles for journals and conferences.

J. Paulo Davim

J. Paulo Davim is a Full Professor at the University of Aveiro, Portugal. He is also distinguished as honorary professor in several universities/colleges/institutes in China, India and Spain. He received his Ph.D. degree in Mechanical Engineering in 1997, M.Sc. degree in Mechanical Engineering (materials and manufacturing processes) in 1991, Mechanical Engineering degree (5 years) in 1986 from the University of Porto (FEUP), the Aggregate title (Full Habilitation) from the University of Coimbra in 2005 and the D.Sc. (Higher Doctorate) from London Metropolitan University in 2013. He is a Senior Chartered Engineer by the Portuguese Institution of Engineers with an MBA and Specialist titles in Engineering and Industrial Management as well as in Metrology. He is also Eur Ing by FEANI-Brussels and Fellow (FIET) of IET-London. He has more than 35 years of teaching and research experience in Manufacturing, Materials, Mechanical and Industrial Engineering, with special emphasis in Machining & Tribology. He has also an interest in Management, Engineering Education, and Higher Education for Sustainability. He has supervised large numbers of postdoc, Ph.D. and master's students as well as has coordinated and participated in several financed research projects. He has received several scientific awards and honors. He has worked as an evaluator of projects for the European Research Council (ERC) and other international research agencies as well as examiner of Ph.D. thesis for many universities in different countries. He is the Editor-in-Chief of several international journals, editor/guest editor of books and journals, editor of a book series, and scientific advisory for many international journals and conferences. Presently, he is an editorial board member of 30 international journals and acts as a reviewer for more than 100 prestigious Web of Science journals. In addition, he has also published as editor (and coeditor) more than 200 books and as author (and coauthor) more than 15 books, 100 book chapters, and 500 articles in journals and conferences (more than 300 articles in journals indexed in the core collection of Web of Science/h-index 61+/12500+ citations, SCOPUS/h-index 66+/15500+ citations, Google Scholar/h-index 85+/25500+ citations). He has been listed among the World's Top 2% Scientists by Stanford University.

Preface

Looking to provide a channel of communication between academics/researchers and managers to disseminate the knowledge of how to manage in a green and responsible environment, the subjects developed in this book assume a particular emphasis in present organizations. Concepts like sustainability, green production/manufacturing, engineering management, green and lean management, social responsibility, high productivity are only some, among many others, about which organizations and researchers are concerned, requiring a break in traditional management paradigms. Indeed, present organizations need to be managed by different models where innovation and change are key words as they drive the organization to success. At this level, and in a manufacturing environment, green production and management appears as a new way to manage, think, and understand organizations, making them more strategic and competitive in the markets where they are or which they seek to occupy in the near future.

The present book, organized in nine chapters, can be used by academics, researchers, human resources managers, managers, engineers, and other professionals involved in green production engineering and management.

Chapter 1 presents "Systems thinking and a value-based definition of sustainability as base for a social enterprise"*;* Chapter 2 contains information about "Blockchain technology and socially sustainable supply chains—A valuation perspective." Chapter 3 discusses "Project selection for sustainable additive manufacturing: A case study," while Chapter 4 asks "Does priority sector lending affect profitability? An empirical study on Indian public sector banks." Chapter 5 studies "The *momentum* for green human resources' management," while Chapter 6 deals with the "Factors involved in using plastic money in the Jangalmahal region of West Bengal: An empirical study." Chapter 7 deals with the "Paradigms of wisdom economy for manufacturing in a post-pandemic world," and Chapter 8 focuses on the "Desertec initiative in Algeria: Challenges and barriers." Finally, Chapter 9 discusses "Connecting ecological economics, green management, sustainable development, and circular economy: Corporate social responsibility as the synthetic vector."

Understood as a critical tool for all professionals in matters related to *green production engineering and management*, interest in this subject is evident for many types of organizations around the world.

The editors are grateful to Elsevier for this opportunity and for their professional support. Finally, we would like to thank all chapter authors for their enthusiasm and availability to work on this project.

Carolina Machado
School of Economics and Management, University of Minho, Braga, Portugal
J. Paulo Davim
Department of Mechanical Engineering, University of Aveiro, Aveiro, Portugal

CHAPTER 1

Systems thinking and a value-based definition of sustainability as base for a social enterprise

Deane Harder[a], **Jan Frecè**[a], **Michail Kyriazopoulos**[b], **and Sauro Bianchi**[b]
[a]Bern University of Applied Sciences, Business School, Bern, Switzerland
[b]Bern University of Applied Sciences, Department of Architecture, Wood and Civil Engineering, Biel, Switzerland

1.1 Introduction

Sustainable development is high up on the agenda of many policymakers. On the ground level, there are a large number of people who would like to make a change or at least contribute. There are, however, numerous practical obstacles in setting up a company with an innovative product and sustainability as a core value. Taking this approach to a higher level, shifting the predominant economic system is even more daunting. This chapter illustrates some of the challenges of founding a social enterprise based on an innovative and sustainable product, derived from research in Switzerland and the Philippines.

Climate change will act as a disruptive force in a plethora of food production systems. A suitable preparation is making the socioeconomic system at risk more resilient. Some initiatives subsequently focus on the sustainable sourcing and the social impact of production. Other initiatives recognize the need for diversifying business models as a contribution to resilience. What has often been neglected is an integrated approach to resilience that exceeds farming practices and considers the whole value chain. To secure positive long-term effects for multiple stakeholders, keystone structural processes, sources of renewal and reformation, and functional diversity need to

Green Production Engineering and Management
https://doi.org/10.1016/B978-0-12-821238-7.00011-7

be optimized. The social enterprise *Cofibo,*[a] therefore, addresses resilience from an interdisciplinary point of view in order to ensure optimum synergies and networking options. The Philippines is the second-largest producer of coconuts. Also, the industry has a high incidence of poverty among coconut farmers making them particularly vulnerable to the effects of climate change. Enabling coconut farmers to adjust and expand existing value chains can generate considerable social and economic benefits. Moreover, understanding the social, economic, and agroecological characteristics of the whole value chain can help to design better agricultural systems. *Cofibo* strives to bridge the gap between research and real-life impact with the development of social and sustainable innovation for the Philippine market and making communities more resilient in the face of potentially disruptive changes.

The coconut industry has a high incidence of poverty among coconut farmers (Yap, 2013), and these smallholder farms are particularly vulnerable to climate change (Intergovernmental Panel on Climate Change, 2014). Coconut is a very versatile tree and yields a variety of products. One part of the tree that is underutilized is the husks of coconuts. These are often discarded or used as compost or firing material (Magat, 1990). Alternatively, they may be used as fibers for ropes, geotextiles, doormats, or baskets, and coir dust can be used as organic fertilizer, rooting media, or pressed into horticultural pots Philippine Coconut Authority (PCA), 2013. The overall vision of this project is to make coconut farming communities more resilient to climate risks by extending their coconut value chains with the applications of the cocoboard technology to mitigate potential disruptive forces like the sudden loss of crops caused by the increasing occurrence of typhoons in the Philippines or other events associated with climate change (Intergovernmental Panel on Climate Change, 2014).

By determining the area of business activity and the approach taken, *Cofibo* has picked a stakeholder segment they are not only addressing with their business activities but also aiming to get their "license to operate" from. To not merely get access to a market but actually gain a foothold in it, a corporation has to be perceived as useful or beneficial by at least one group of stakeholders that

- has access to the corporation,
- has the means to take advantage of this access, and
- is motivated to do so.

[a]Name of the company changed.

In this paper, we describe the setup of a social enterprise with a systems thinking mindset. We argue how a shift in economic thinking from linear to circular is driving a change in the way business is done and seen. Consequently, this influences the way businesses are set up, even more so, when considering a social enterprise. Adopting a more circular notion of economy may provide a suitable alternative; however, it requires some rather unconventional ways of thinking in the design phase of setting up the business. We show how a systems thinking approach can help in defining a company and preparing corporate development and operations management. A particular emphasis is placed on the role of functional corporate values. We explain what we mean by this term and how corporate values are linked to other factors and stakeholders in particular. Finally, we discuss how functionally formulated corporate values predetermine the boundary conditions for corporate activity and development and thereby constitute an important guidance for stakeholders even in the early stages of a startup corporation and at the start of production. The article is concluded with some reflections on the way a social enterprise can navigate the business context and how a thoughtful design based on systems thinking can minimize the risk of unintended consequences that may counteract the good intentions at the start of a social enterprise.

1.2 A shift in economic thinking

Our current predominant economic system is directly or indirectly based on neoclassical theory of economic growth (Goodland & Ledec, 1987). One of the de facto paradigms for many entrepreneurs is that companies exclusively exist to make profit (cf. Friedman doctrine) (Friedman, 1970). This mindset is more and more questioned, in particular regarding a more sustainable development (Frecè & Harder, 2018) and with social enterprises (Austin, Stevenson, & Wei-Skillern, 2012). Instead of solely focusing on shareholders as propagated by the Friedman doctrine, social businesses operate with the much broader and more complex term stakeholder, as introduced in the 1980s (Freeman, 1984). By recognizing their shareholders and their perspective on the world, a social business willing to question traditional approaches and best practice processes can innovate in a way supporting its stakeholders, e.g., by offering them alternative business approaches (e.g., circular business models) or by entirely leaving the realm of the established and proceed with custom-tailored, academia-supported innovation.

The starting point of the value proposition of the social enterprise *Cofibo* is the so-called cocoboard technology, the outcome of a joint R&D project between Switzerland and the Philippines (2014–2017). It combines coconut husks and plant-derived tannin-based adhesives to a composite particle board comparable to wooden particle boards. From a technological and socioeconomic perspective, there are three objectives that *Cofibo* would like to achieve as a social enterprise:

1. Designing a social franchise system with a blueprint for a scalable production process and supply chain management for raw materials for the cocoboard technology as keystone structuring processes for coconut farming communities as franchise holders.
2. Setting up a "Design Lab" to integrate and accommodate for new applications and innovations of the cocoboard technology to create a source for renewal and reformation with relevant customers and representatives of the franchise holders.
3. Managing the value chain of cocoboard technology with an affiliate social enterprise and enabling and supporting policymaking to ensure functional diversity and providing governance to the franchise system.

Besides the "social" aspect of the enterprise, *Cofibo* also has a corporate value of sustainability. It is, therefore, necessary to integrate the concept of sustainability in the value system and, subsequently, business operations of the company, as will be detailed in Chapter 3. To this avail, we will first look into this chapter at the concept of corporate sustainability, followed by some definitions of social entrepreneurship and innovation. Then, we will explain a potential paradigm shift from linear to circular economy and indicate how academic research may act as a leverage point for innovation.

1.2.1 Sustainability as a value-based concept

Sustainability is an abstract concept. As such, it requires models and visualizations to make the concept more tangible and comprehensible. One model of sustainability is to consider nested systems, e.g., the economic system as a subsystem of the social system of society, which in turn is a subsystem of the encompassing supersystem of nature (Fig. 1.1). Thus, any change in the interconnection of factors or their dynamics will have more or less strong rippling effects in the connected systems (McKelvey, 2002). As there are many definitions of sustainability proposed, we deem it useful to emphasize the practical implications. Therefore, the way sustainability is framed for this article is *a balanced and systemic integration of intra- and intergenerational economic, social, and environmental performance* (Geissdoerfer,

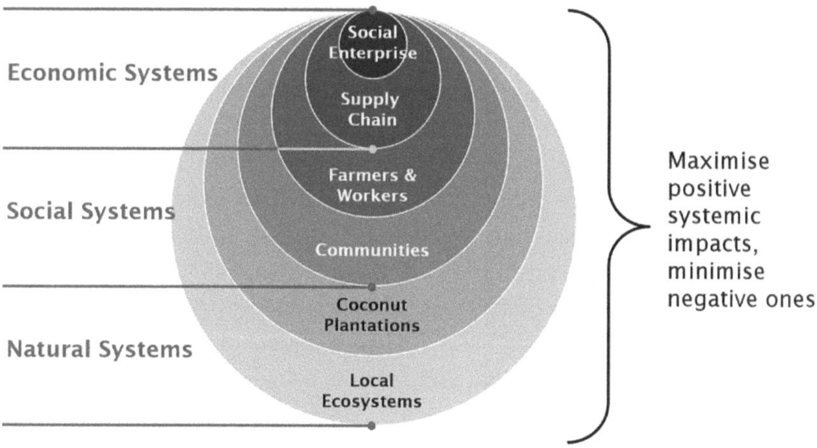

Fig. 1.1 Sustainability of a social enterprise from a systems thinking perspective.

Savaget, Bocken, & Hultink, 2017). As this is from a societal perspective, we also used the definition from a corporate perspective (Frecè & Harder, 2018): *sustainability entails that the future is a better, healthier place than the present* (Blewitt, 2008). The specific operationalization of what is considered "better" or "healthier" is dependent on the values and perceptions of the company and its stakeholders (Christen, 2010).

Applied to the *Cofibo* project, an onion-like structure of layers within the systems and subsystems can be identified. The most inner layer is the actual company, embedded by its supply chain as essential parts of the economic subsystem. This is nested in the social systems of the farmers and workers that are linked to the company and its supply chain. The farmers and workers are part of communities, which form the outer layer of the social subsystem. This subsystem is nested in the natural supersystem, which consists of layers of specific coconut plantations that are part of the surrounding local ecosystems. As a model, this helps to make some of the interdependencies more plausible but also illustrates how reality is grossly simplified to make certain aspects more manageable.

1.2.2 Social entrepreneurship and innovation

Managing a company is inseparably linked to entrepreneurship. To avoid conceptual ambiguities, the terms entrepreneur and social entrepreneurship will be briefly explained. The term entrepreneurship is used according to the definition of Fritsch (Fritsch, 2019) (translated by authors): "Entrepreneurship means pro-active undertaking in the sense of founding and managing a company. Successful entrepreneurship requires the recognition of entrepreneurial opportunities, creativity, initiative, the will to shape things, personal responsibility, assertiveness, and the willingness and ability to take risks." The term social entrepreneurship, on the other hand, is more difficult to define. Accordingly, Dacin et al. count a total of 37 definitions in the area of social entrepreneurship/social entrepreneur (Dacin, Dacin, & Matear, 2010). To keep the discussion as practical as possible, we will apply the definition of social entrepreneur in the sense of Dees (2018) and a supplementary definition of social entrepreneurship by Austin et al. (2012): "Social entrepreneurs play the role of change agents in the social sector, by:

- Adopting a mission to create and sustain social value (not just private value),
- Recognizing and relentlessly pursuing new opportunities to serve that mission,
- Engaging in a process of continuous innovation, adaptation, and learning,
- Acting boldly without being limited by resources currently in hand, and
- Exhibiting a heightened sense of accountability to the constituencies served and for
- the outcomes created."

In addition, social entrepreneurship should be understood as "innovative, social value creating activity that can occur within or across the nonprofit, business, or government sectors" (Austin et al., 2012). For a better delimitation, it should be mentioned that social entrepreneurship does not represent a sector of its own, should not be understood synonymously with social business, and, in contrast to CSR, always includes the elements of entrepreneurship and innovation.

An innovation can be defined as a novelty that confers an economic advantage (Gassmann & Sutter, 2013). This definition emphasizes that not every new product or service on the market automatically constitutes an innovation. At the same time, this definition also covers innovations in the area of organization, processes, or business model. In this context, social innovation should be understood as an innovation that perceptibly improves a concrete problem in society that has been identified as such (Huybrechts & Nicholls, 2012). A social innovation is, therefore, less about the economic benefits than about the social impact. Accordingly, a market-based approach is not assumed, but is possible. Such social innovations can manifest themselves in different ways, e.g., as grassroots movement, association, citizens' initiative, or—as with our case at hand—a social enterprise. Since the focus here should be on Green Production Engineering and Management, social innovations in the form of enterprises will be discussed in the following chapters, as an interface between social innovation and market-based social entrepreneurship.

1.2.3 From linear to circular economy

A concept often related to sustainability is circular economy. Although these two concepts are commonly linked, there are clear intellectual boundaries (Geissdoerfer et al., 2017). The conceptual foundation can be found in ideas like cradle-to-cradle (McDonough & Braungart, 2010), industrial ecology (Graedel & Allenby, 2003), looped and performance economy (Stahel, 2016), or blue economy (Pauli, 2010). As a simple definition of circular economy, the framing of the Ellen MacArthur Foundation is adopted for this article as an *industrial economy that is restorative or regenerative by intention and design* (Ellen MacArthur Foundation, 2017). In other words, circular economy is a practical concept to implement sustainability based on specific corporate values (Frecè & Harder, 2018).

The starting point for implementing a business based on circular economy is considerations on sustainability. Following the line of arguments of Tukker et al. (2016), there are three main areas on an individual level that define the degree of sustainability:

1. What we eat (food)
2. How we move around (mobility)
3. How we build our houses and set up our lives (housing)

Overall, to maintain favorable living conditions for humans, many experts see the necessity to develop efficient, consistent, and sufficient innovations for resource decoupling and impact decoupling (Fig. 1.2) (UN Environment: International Resource Panel, 2017). Moreover, there seems to be an understanding among a growing number of economists that the existing economic model will not generate enough change in time to provide viable solutions and alternative models are seen as more promising, e.g., models of circular economy. *Cofibo* is an example of how an improvement in general well-being and the generation of economic activity does not need to lead to undue high amounts of resource use or negative environmental impacts. It uses by-products of existing food production (coconut husks) and aims to utilize local supply chains (mobility) to provide building materials for interior, non-loadbearing uses (housing). The value chain of *Cofibo,* therefore, touches upon all three relevant areas that need sustainable innovations. It is, therefore, a prime example of a low-tech innovation in line with the requirements of a circular economy.

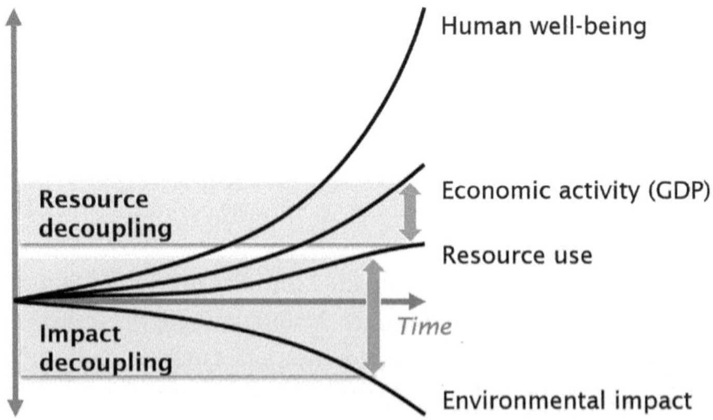

Fig. 1.2 Decoupling natural resource use, environmental impact, and economic growth. *Adapted from UN Environment: International Resource Panel. Decoupling natural resource use and environmental impacts from economic growth. In: [Internet]. 3 Feb 2017 [cited 14 Jan 2020]. Available: https://www.resourcepanel.org/reports/decoupling-natural-resource-use-and-environmental-impacts-economic-growth.*

A circular economy is organized in loops. Rather than the conventional linear model of "take, make, dispose," a circular economy strives to decelerate, reduce, reuse, and recycle all materials and maintain value as long in the value chain as possible (Kalmykova, Sadagopan, & Rosado, 2018). A simple model can be defined with seven steps along the value chain (Fig. 1.3):

1. Materials sourcing
2. (Re-)design
3. Production
4. Distribution and sales
5. Consumption and use
6. Recycling and recovering

This simple model already indicates that the core of circular economy is not just eco-efficiency—which can also be achieved in a linear economy—but rather involves all stages of the value chain and acts in value and material loops in between the stages. For the social enterprise considered here, the focus will be on materials sourcing and recycling. The materials are coconut husks as a by-product that otherwise would be devalued as compost. The glue to make particle boards can be either based on petrochemical adhesives or, alternatively, natural tannins sourced from local plants. Although the natural tannins are favorable from a sustainability point of view (sourcing and disposal but also making them free of toxic or allergenic emissions during use), the technical feasibility and production processes have to be taken into

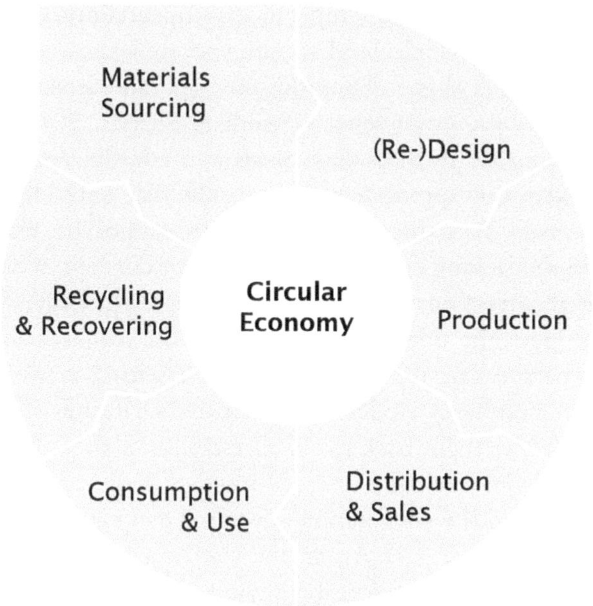

Fig. 1.3 A simple model of circular economy. *Adapted from Kalmykova, Y., Sadagopan, M., Rosado, L. (2018). Circular economy—From review of theories and practices to development of implementation tools.* Resources, Conservation and Recycling, 135, *190–201. https://doi.org/10.1016/j.resconrec.2017.10.034.*

account and current quality requirements for a commodity market like particle boards have to be met. Potentially, a start of production with coconut husks but a conventional glue might be recommended; however, the design of the particle boards allows for a later switch to a produce an all-natural product with no toxic emissions. The distribution and sales are conceptualized in a franchise system to ensure local production, delivery to the market, and use rather than covering long distances or importing materials. The use as interior, non-loadbearing building material is foreseen to provide cost-efficient and high-quality particle boards, comparable to wood-based particle boards. The recycling is planned as a complete return of all materials to the natural sphere as the intended use of only natural raw materials makes degradation at the end of life of the particle boards easy to implement. The example company of *Cofibo,* therefore, takes an integrated approach to circular economy that considers every stage of the whole value chain.

Lacking a clear definition of corporate sustainability is a major risk for social enterprises like *Cofibo*. Not being aware of what impact corporate activity has on sustainability and what one's contribution is in this regard endangers alternative business setups. In case uncertainty arises in regard to business decisions, people tend to cling to well-known and internalized value assessments. Vague definitions of corporate sustainability lead to lacking or equally vague guidelines (Smith & Sharicz, 2011). Unless the employees take matter in their own hands in a coordinated effort, which is highly unlikely, this approach ends in results that only loosely fit with the original goals. This discrepancy, however, cannot be exactly determined due to the lacking corporate sustainability concept, which prevents target-oriented corrections but rather contributes to the general perception of vagueness. Here, the thinking process closes the feedback loop. From a stakeholder's perspective, the vagueness in goal setting as well as evaluation of goal achievement erodes the alternative character of the business initiative and throws them back to a well-established, more financially and less socially or environmentally oriented mindset. In a negative reinforcing cycle, this reflects on the corporation's decision-making and reinforces the trend even further (Fig. 1.4). Again, the thinking process is caught in a feedback loop. It should, therefore, be obvious that a convincing setup of a business in a circular way requires systems thinking and a clear definition of corporate sustainability.

We propose to base the concept of sustainability on the broad premise that "the future is a better, healthier place than the present." This premise should then be further detailed and ethically profiled by functionally formulated corporate values. These corporate values and the vision of a "better

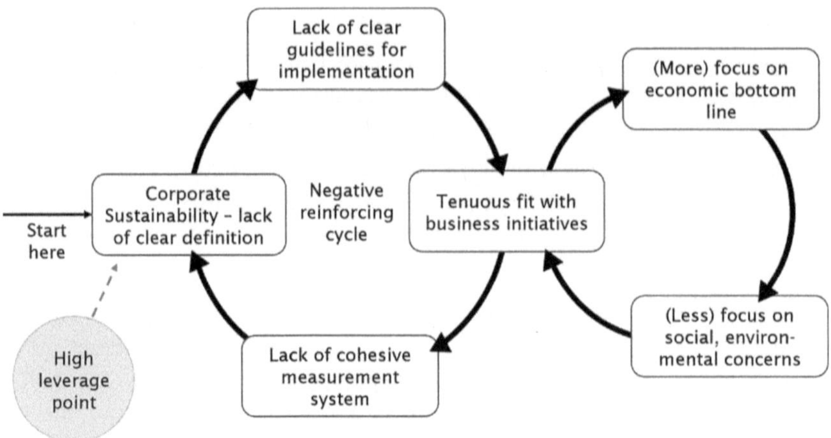

Fig. 1.4 The issue of having an unclear concept of corporate sustainability. *Adapted from Smith, P A, & Sharicz, C (2011). The shift needed for sustainability. Learning Organization, 18(1), 73–86..*

place" and "doing good" guide the efforts in corporate sustainability toward contribution, restitution, and/or compensation, accommodating the fact that most business activities cannot avoid harmful impacts overall (Frecè & Harder, 2018). We will explain the intricacies of a value-based concept of sustainability in Chapter 3.

1.2.4 Academic research as leverage point for innovation

The base of the value proposition of the social enterprise *Cofibo* is the so-called cocoboard technology. This innovative technology was the outcome of a joint R&D project between Switzerland and the Philippines (2014–2017). It combines coconut husks and plant-derived tannins to create an all-natural composite particle board comparable to wooden particle boards. The final product is very promising, but still in the pre-industrial phase (Fig. 1.5). A successful innovation matches product properties with customer needs. In order to do this, significant funds or a low-competitive market environment are required to optimize the product to a competitive level. The project, therefore, aims to commercialize and further develop applications of the cocoboard technology in government-supported or regulated social housing and public buildings.

Tannins are natural compounds that may be sourced through hot-water extraction from bark and other biomass of various tree species. Tannins have proven efficacy as natural substitutes in diverse applications like leather tanning, wood adhesives, biocides, or additives for health care and food. Unexploited biomasses from both the wood and coconut value chains (e.g., barks, roots, and husks) might thus be valorized as tannin sources and represent an additional income opportunity for rural communities in developing

Fig. 1.5 *Cofibo* is based on the previously developed cocoboard technology, making particle boards from coconut husks to provide building materials, e.g., ceiling panels and other interior, non-loadbearing applications.

countries, whose livelihood is strongly connected with the use of local natural resources. *Cofibo* strives to provide a cost-effective, sustainable, and integrated tannin extraction value chain in the Philippines, coordinated through a franchise system.

1.3 Transferring knowledge into action

Resilience can be defined as the amount of disturbance that a living system can withstand without changing self-organizing processes and structures (defined as alternative stable states) (Gunderson, 2000). Key factors for resilience are as follows: (1) keystone structuring processes across a number of scales, (2) sources of renewal and reformation, and (3) functional diversity. The social enterprise focuses on increasing the resilience of coconut farming communities by addressing all three factors. Resilience and sustainability are closely connected. A sustainable value chain requires responsible operational management (OM). Environmental and social responsibility in developing countries, however, is still nascent research areas in operational management (Goodland & Ledec, 1987). As these attributes are at the base of the planned social enterprise, they require profound analysis. Since this information is not available yet, we will refer to standard academic concepts and a system dynamics model for *production starts* (Sterman, 2010) to highlight existing options to plan with a systems thinking mindset. From a scientific point of view, the paper aims at answering four central questions:

1. How should a social enterprise get a societal "license to operate"?
2. What are the key requirements and obstacles in the selected business context?
3. What is the potential of circular business models and its implementation?
4. How can corporate values provide guidance for corporate development?

To address these questions, we will first detail how the social enterprise *Cofibo* intends to have a positive overall impact on sustainability by contributing to economic, ecological and social sustainability, as well as setting itself up as a low-end innovation and social innovation.

1.3.1 Economic sustainability

As a company operating in a conventional market, *Cofibo* has the obligation to act at a competitive business level, however, without the need to generate profit; reaching break-even point is sufficient. In addition to the economic impact of the local material sourcing and the provision of local income opportunities, to increase the productivity and resilience of coconut farming communities, the social enterprise *Cofibo* aims to train farmers in (1) preparation of raw materials and (2) organization of value chains, and to enhance sustainability and resilience (in particular to risks associated with climate change) of coconut farming (International Tropical Timber Organization, 2004).

1.3.2 Ecological sustainability

The sourcing of raw materials for the cocoboard technology has an impact on the ecosystems in which the coconut farmers operate and which forms their main source of livelihood. Thus, an accurate and comprehensive ecological analysis is crucial in order to make the value chain of the cocoboard technology more resilient. Cofibo has the objective to analyze the ecological footprint (EF) and life cycle assessment (LCA) applied not only to the material sourcing but to the entire chain of value as part of its sustainability management.

1.3.3 Social sustainability

Given the approach and setup of the social enterprise *Cofibo,* the aim is to work with people rather than just employing "human resources." The cocoboard technology is to be developed for positive long-term effects and promoting sustainability and resilience among smallholder communities. It is thus mandatory to monitor the social effect of cocoboard production and application over time (e.g., social acceptance, actual or perceived risks associated with the cocoboard technology and its application), which will be part of *Cofibo's* sustainability management.

1.3.4 Low-end innovation

The whole *Cofibo* concept is designed to foster the development of applications that can be produced by a minimum of technological equipment, allowing for a vertical integration locally so that a large part of the value chain can be captured within small local coconut farming communities; a typical low-tech, low-cost development subsumed by the term low-end innovation (Reinhardt, Gurtner, Hoskins, & Griffin, 2017; Segelitz-Karsten, Reinhardt, Hietschold, Gurtner, & Walsh, 2019). To maximize the economic and social impact, it is, therefore, necessary to develop concepts and guidelines for technology transfer, training, and tests in the field (Nawir & Santoso, 2005; Siregar, Rachmi, Massijaya, Ishibashi, & Ando, 2007).

1.3.5 Social innovation

Once the cocoboard technology is successfully transferred to a new production site, the blueprint of the system can be franchised to other regions or countries. The introduction of cocoboard technology is expected to cause positive social and socioeconomic impacts. Likewise, the implementation of the *Cofibo* franchise system can be expected to export the creation of value along the whole value chain in other places as a driver of societal change.

Based on these objectives of the social enterprise *Cofibo,* we will address in the following chapter the conceptualization of a business vision and mission to gain a societal license to operate, before we describe functional values as base for managing a social enterprise.

1.3.6 Vision and mission to gain a societal license to operate

The identity of an organization is driven by its entrepreneurial vision and mission. The vision and mission for social enterprises can be easily derived from the model of strong sustainability (Fig. 1.6, left). The vision should address a concrete need in society for the satisfaction of which the company wants to and can take responsibility. One example could be "There should be more housing for the poor." The reversal, i.e., the 100% fulfillment of this need, then represents the vision, e.g., "Affordable houses for all." The mission represents the way the company takes responsibility for the fulfillment of this vision, e.g., "We provide good-quality and sustainable building materials that helps to make houses affordable." This is a first approximation of the evaluation criteria of the company's main value-added process, i.e., in the example, the criterion

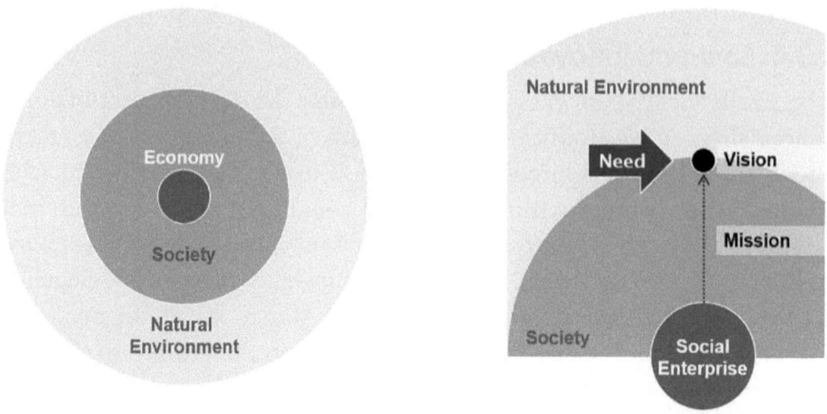

Fig. 1.6 Model of strong sustainability (*left*) and derivation of corporate purpose, vision, and mission (*right*) as the basis for integrated sustainable corporate management.

of whether the building materials produced are meeting the quality requirements and, derived from this, to give recommendations for action for people, who want to build affordable and/or more sustainable houses. The mission and its practical implementation determine whether the social over-system grants the economic subsystem "enterprise" the right to exist and thus issues it a "license to operate." With this simple model, the reason and purpose of a social innovation can therefore be easily derived (Fig. 1.6, right). In social enterprises, this simplicity and clarity of vision and mission are of particular importance to justify the business character and meet the various expectations of internal and external stakeholders; accordingly, these aspects should be presented transparently as a working basis in an integrated sustainability management and expressed or accompanied by functional corporate values.

1.3.7 Functional values as base for managing a social enterprise

We have described the elements of normative management with a simple model to define the vision and mission of a company. In addition, corporate values need to be set. In this chapter, we will characterize functional corporate values, discuss value dysfunctionality, and describe sustainability as a value-based concept.

Transferring theoretical knowledge into action means striking a balance between multicritical interdependencies. This may happen in various ways, depending on the local preconditions regarding a plethora of factors like available infrastructure, local terrain, level of knowledge, social structure, local traditions, and individual and/or social preconceptions. The traditional way to address this challenge in a franchising context is a network of role descriptions, procedural guidelines, provisions regarding used equipment and material sourcing, etc. (Paswan & Wittmann, 2009). The primary goal of such measures is a high level of standardization with regard to the end product sold to the customer. The equally standardized production process is in this perspective merely a supporting goal, minimizing the probability of end-product divergence.

In the case of *Cofibo,* however, the defined franchising goal is not primarily a minimized production divergence, but rather a minimal divergence regarding the focus on

- increasing long-term productivity and resilience of producing communities and their value chains
- enhancing the resilience against fluctuations in harvest and other risks related to climate change

- ensuring that sourcing of raw material is based on sound assessment regarding its environmental and social impact
- achieving the goal using as low-level technology as possible
- gathering and passing on experiences and lessons learned (knowledge sharing)
- a generally participatory approach respecting and incorporating the community's needs and capacities

Especially, the latter points make it fairly clear why the traditional franchising approach with its strict, explicit regulations regarding each step of raw material sourcing, production process, and delivery of the end product cannot meet *Cofibo's* needs (Tsay, Nahmias, & Agrawal, 1999). A participatory approach is the opposite of top-down work instructions and thus requires a quite different organizational focus, one based on organizational or maybe even "franchisal" values like trust (Richard, Frazer, Weaven, & Dant, 2016).

1.3.7.1 Values vs business targets

Corporate values are easily confused with one of the many targets a company sets itself. In contrast to a business target, e.g., achieving a customer satisfaction rating of over 85%, corporate values cannot be simply achieved by a certain point in time. They are either part of the corporate identity, the corporation's way of doing business, and thereby—if done right—also part of its employees', or they are not. Corporate values are not goals one wishes to achieve sometime, nothing to strive for in the future, but they are either lived and expressed in daily business activities, or they are not. Thus, corporate values are not a free choice but rather the description of typical organizational characteristics, comparable to the personality of an individual. No individual would identify as Buddhist because they sometime in the future work plan to work toward the core values of Buddhism.

Both business targets and corporate values are ways to distinguish right actions from wrong ones. However, while business targets are defined target states that can be reached, ticked off, and replaced by a new target, corporate values are a way of doing things not because this behavior leads to a predetermined target state, but rather because the behavior itself is considered the target, regardless of whether this leads to a concrete business target or not. At this point, one could misunderstand targets as procedural instructions or a code of conduct, distinguishing wanted behaviors from unwanted ones. A corporate value, however, is a more complex instruction, not simply aiming at triggering the same behavior regardless of the context. If corporate values demand a focus on the long-term resilience and productivity of the communities involved, the communities are never too resilient or too productive to not live that corporate value every day. Naturally, it is still possible that any particular corporate value may not be the primary focus in communities, which have already reached a relatively high and comfortable level of resilience and productivity. But even in these cases, it never ceases to be a valid corporate value, making any concrete actions against the productivity and resilience of the community at least questionable, as they contradict active values the corporation holds. Therefore, corporate values are at the same time guidance and ideals.

1.3.7.2 Value dysfunctionality

There are a range of reasons for corporate values to become dysfunctional. The two discussed below are selected for three reasons:

1. Failing to achieve a proper definition as well as failing to cover value interplay can be considered the two most common and widely observed mistakes in the way corporations deal with corporate values.
2. In a traditional corporation, the ever-present dogma of growth and profit maximization creates a basic common understanding of purpose and direction. Enterprises like *Cofibo* are lacking such a well-established, cross-culturally, and time-proven dogma. They, therefore, have to be especially clear and unambiguous when creating a common perspective, sense of purpose, and understanding of acceptable modes, means, and ends.
3. The level of ethical integrity and consistent corporate behavior expected by relevant stakeholders is particularly high with social enterprises.

Insufficient definition

Corporate values are in some way letters of intent, and, if those intentions are not well-defined, they leave leeway for individual interpretations. This in turn leads to uncertainties regarding the identification of stakeholders with the corporation and its values. How can they fully identify with a corporation if they are not sure what it is about at its core? It is, therefore, crucial for an operation like *Cofibo* to clearly state for each of their corporate values the basic information of

- how they relate to the corporation's mission and vision,
- what their contribution is in this regard,
- within which timeframe such contributions are expected to noticeably become apparent,

- how these contributions are going to be measured,
- who is affected by any particular corporate value in which way, and
- on whose initiative and through which process any particular value became part of the corporate identity.

While a certain amount of vagueness regarding such "details" of the definition of corporate values might be attractive from a marketing perspective, as it allows for an easier and broader inclusion of people and their views, it bears serious risks. If this rather superficial approach is taken, common understanding may turn into ideological fighting during the implementation phase. Because corporate values were not clarified beforehand, the potential damage the corporation could suffer from regarding their finances, plans, market goodwill, and ideological integrity could be considerable.

Undefined value interplay

Establishing a network of corporate values without at least one corporate value contradicting another one is a conceptual constellation nobody should count on. Rather, such value interplay has to be foreseen (or at least detected when occurring for the first time) and handled on a generic level. In the case of *Cofibo,* value interplay is already clearly foreseeable in the example of the two following values:

- Increasing long-term productivity and resilience of producing communities and their value chains.
- Ensuring that sourcing of raw material is based on sound assessment regarding its environmental and social impact.

In the not unlikely scenario that processing raw materials without sound assessment regarding environmental and social impact is the easiest way to improve local resilience and productivity, to what degree will concessions be made? To what degree must resilience and productivity be endangered, e.g., because raw materials with a sound assessment currently cannot be delivered reliably or in needed amounts, before compromises are made?

Investing a high level of attention in the complex issue of value interplay—the above-mentioned example being merely one of many value interplays worth a more in-depth discussion for *Cofibo*—ensures a high functionality of corporate values and provides the foundation for a stable relationship with all stakeholders addressed by these values. Only if properly informed, stakeholders can know that the corporation will prioritize certain values over other ones if push comes to shove. Only then will they not be unpleasantly surprised to see the values that made an identification with the corporation possible in the first place, being ignored in favor of other corporate values with less subjectively perceived importance to them. Functional values are crucial for any form of organization in order to avoid potentially disruptive uncertainties and misunderstandings, the perception of volatility or the inability to build lasting trust toward its stakeholders (Frecè, 2019). As already mentioned, this is especially true with social enterprises, operating outside the traditional dogma of growth and profits first.

1.3.7.3 Sustainability as a value-based concept

It can safely be assumed that an alignment of corporate values with consumer ones will have a conducive effect on doing business. If corporate values are coherently operationalized, then the offered product and value chain will be perceived more positively than with contradicting or misaligned values—independent of the actual problem-solution fit of the product (Horlings, 2015). In a first normative approach, it is, therefore, necessary to assess the potential positive and negative impacts of the business activities and manage the associated values, effects, and consequences. Laid out as a high-level 5-step process (Fig. 1.6), value assessment can be defined by:

1. Premise: stating the contribution to sustainability, defined by "the future is expected to be a better, healthier place than the present."
2. License to operate: defining the relevant values, goals, and issues of the company.
3. Self-characterization: describing the sustainability profile of the company and associated functional corporate values.
4. Reflection: assessing the resources that are depleted or devalued by the business activities of the company.
5. Integration: setting the standards and goals for efforts to restitute or compensate negative impacts caused directly and indirectly by the company.

As every corporation has some kind of impact on sustainability factors and no management is able to control impact at 100% on every level, corporate sustainability becomes a question of corporate prioritization and valuation. The question, whether lowering wages in order to have the financial resources to improve the environmental impact of the production process or the health impact of the product itself, cannot be answered universally or generically, as it lies in the eye of the beholder or rather in an individual's prioritization of values. Consequently, the goal of a functional value declaration is not to present the best, most broadly likable image of the corporation, but rather to present its prioritization and valuation mechanisms and thereby attract stakeholders matching their world view and their pursuit of sustainability.

As corporate values are supposed to guide corporate actions toward a desired goal, they must be subject to control mechanisms ensuring that the guiding taking place does indeed bring the corporate closer to the stated goal. Such control mechanisms need to ensure functionality on several levels: value issue, value goal, value activity, and value norm. Finally, two perspectives need to be considered: the corporate perspective and the stakeholder perspective on values.

Value issue

Value issue describes the driving idea or desire behind the value-based activities and addresses the question of the importance of a topic and the stance of the corporation on it. In contrast to the two following levels, the value issue solely depends on the value systems of the value declarators and their vision for the corporation's identity and, therefore, values. There is no logically wrong statement in a value issue as it states personal preferences and perceptions of importance. Typical examples for a value issue would be, e.g., climate change, declining corporate profits, or decreasing religiosity to name a few, addressing a broad scope of ideals. Value issues are often formulated in a corporation's vision statement, which announces to the public why this corporation exists and what it considers to be the conditions it no longer needs to exist. To illustrate this last point, consider, e.g., a corporation striving for a world where no clothing contains any chemicals, potentially dangerous to humans or the environment. Such a company would need a new vision with new value issues as soon as the planet was free of any chemically treated clothing, whenever this may be the case. Thus, the value issue should capture the core of the business.

Value goal

Value goals describe what objectives the corporation plans to achieve in order to support the value issues defined in the preceding step. Therefore, such goals have to be, at least in theory, achievable within a finite timeframe. Furthermore, it has to become clear, in which regard value goals support at least one the formulated value issues. As a consequence, there is an inevitable functional dependence between value issues and value goals. Value goals are typically formulated in a corporation's mission statement. There, a corporation states publicly, which goals it is going to pursue to achieve the desired state described in the vision statement and why these goals have the potential to get society and/or the company closer to the envisioned target state. By reading the mission statement and understanding the underlying value goals, stakeholders should be able to understand the corporation's prioritization and valuation mechanisms. As an illustration, take again the example of a corporation striving for a world where no clothing contains any chemicals potentially dangerous to humans or the environment. This corporation could strive toward this goal in a number of different ways. It could address the problem by

- producing clothing free of chemicals
- developing and promoting new technologies, making chemicals obsolete in the clothing production process
- developing and promoting post-production techniques to clean the clothing of chemicals, etc.

Only when the stakeholders understand which path the corporation intends to pursue and why, they get a better impression whether the corporate values match their personal ones, thus granting a value-based support and, if there are sufficient supporting stakeholders, contributing to a societal license to operate.

Value activity

Value activities incorporate all activities that are to be performed in order to reach the value goals determined before. As the corporation as a whole exists to pursue the value goals, no corporate activity should take place without being rooted in a corporate value goal and thereby in a corporate value issue. Value activities represent the actual implementation of a corporate value, not just the code calling for it, and are, therefore, the often neglected, heart-piece of corporate value application (Webley, 1999). As value activities are to lead toward a goal, they must be functionally connected to the value goals they are expected to advance. However, value activities do not need to be directly linked to the level of value issues as they depend on a functionally intact connection between value goals and value issues. Properly and accurately performed, value activities can, therefore, only have the expected influence on the value issue if the intermediate value goals have been determined correctly. Again, let us take the corporation striving for a world where no clothing contains any chemicals potentially dangerous to humans or the environment decides to achieve this vision by developing and promoting new technologies making chemicals obsolete in the clothing production process. Assuming no other corporate values are in place leading to value interplay, a stringent value activity could be to mass-produce the new technology in low-wage, low-standard countries in order to keep the price of the end product low and thereby make it accessible for the highest possible number of textile producers. Value activities thus explain the doing on a corporate level.

Value norm

Value norms fulfill a dual function. For stakeholders, they represent an indicator for the extent the corporation is planning to implement a value activity. They can use this to match the radicalness of their own values against the corporation's. For the corporation, value norms are used to determine when a value activity can be deemed as satisfactorily completed. It should, therefore, be formulated in an interpersonally measurable form. Fulfilling a value norm, however, can only result in a successful activity if the link between activity and norm has been defined in a target-oriented way. In the case of our example corporation, pursuing the value activity of keeping the price of their end product low and the product thereby widely accessible can for example result in setting the target production price to $15 a piece. This norm provides an indication for stakeholders, how the corporation is planning to operate and to which extent they are willing to ignore their negative social and environmental impact during production to keep the production price at a certain level. It provides the stakeholders with a foundation to decide whether their personal values prioritize better conditions during production or whether the broad accessibility of the end product outweighs the harm done. Value norms thus provide a form of internal and external calibration.

Two perspectives on values

As corporate values function as an important way of communication between corporation and stakeholders, there are two perspectives to discuss:

- From a corporation's perspective, corporate values are needed to establish the corporate view on sustainability and its implementation. Once established, guidelines can be set to ensure corporate actions reflect corporate values and the ways and means how the corporation contributes to sustainability can be defined. From that point on, the results of corporate actions have to be assessed and corrections made if the results do not match the expectations.
- From a stakeholder's perspective, the corporation's contribution to sustainability can be assessed and matched against personal expectations in this regard. Depending on the result of this matching process, stakeholders decide to continue granting this corporation a license to operate, turn elsewhere, or in extreme cases turn against the corporation (Fig. 1.7).

The usefulness of corporate values, overall, is often underestimated. If they are functional, they serve multiple purposes and facilitate the interaction

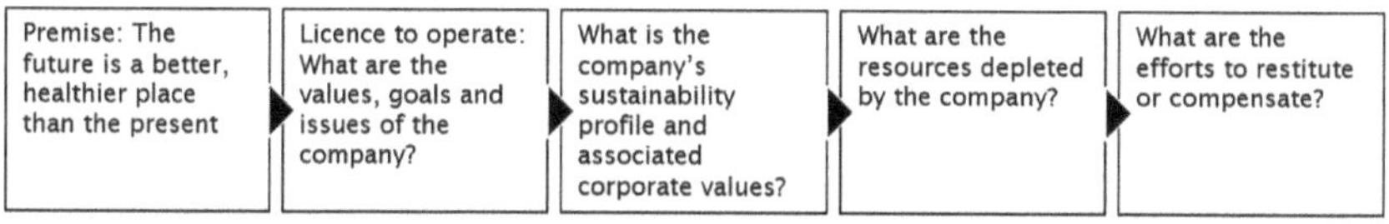

Fig. 1.7 Integrating sustainability in the DNA of a company.

with stakeholders. Even more importantly, they make sustainability a tangible concept for a company. Subsequently, the careful and well-reasoned definition of corporate values should be an essential step in the setup of a social enterprise. In the next chapter, we outline how a social enterprise can be managed with a systems thinking mindset, based on functional corporate values (Fig. 1.8).

1.4 Managing the social enterprise

In the beginning of a social enterprise, there is usually a vision. Imagining an ideal state is then followed by first suggestions on how to achieve this state. Once the why, what and how (crucial part of corporate vision and mission) are taking shape, functional corporate values should be defined, expanding the corporation's vision and mission statement. Details in the business model need to be filled in and there are plenty of useful tools around to do so, e.g., the Business Model Canvas (Osterwalder & Pigneur, 2010) or templates for business plans. Here, we would like to focus on linking the setup of a social enterprise with subsequent corporate development,

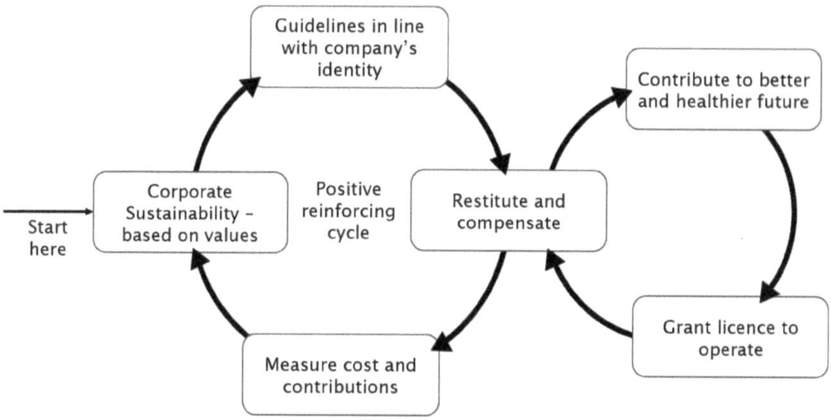

Fig. 1.8 Doing business by operationalizing corporate values in a coherent manner.

stakeholder management, and production start. Finally, we take a look at how a social enterprise may navigate the business context.

1.4.1 Corporate development from the start

Organizations can be conceived as social systems (Harder & Tokarski, 2018). As such, there are certain typical features that characterize organizations. Some of the most salient ones are as follows (Simon, 2004):

- Once created, there is a tendency for self-sustaining (autopoiesis).
- Once a first dynamic equilibrium of various factors is reached, inertia develops, perpetuating this state—if possible (homoiostasis). Subsequently, it takes strong impulses (irritations) to change the system.
- The structural elements are formed by the communicative couplings of the actors within the system. These develop spontaneously and more or less randomly (emergent).
- The communicative couplings generate interaction patterns that define large parts of the corporate culture (the "way we do things here").

Corporate development aims to make organizations more responsive and adaptable in finding solutions to business challenges while increasing their resilience in the face of changing external and internal conditions. These conditions can be particularly challenging for a social enterprise. With high expectations from internal and external stakeholders and a competitive market, leadership needs to account for and integrate multiple aspects. Especially, the setup phase of a company is often characterized by a rapid succession of changes. As many changes in one organizational element entail unintended consequences in other elements or over time, it is, therefore, useful to consider corporate development right from the start of a social enterprise (Fig. 1.9).

If an organization can be considered a social system, corporate development is the attempt to encourage changes within the system in a thoughtful and intentional way. There are various models to structure corporate development. For this article, a model by Glasl and Kalcher (2014) is used (Fig. 1.6). One key aspect of this model is a description of the system with seven basic elements, namely identification, strategy, organizational structure, groups and leadership style, roles and functions, processes, and infrastructure and materials. The individual elements are described in more detail in the following.

1.4.1.1 Identity

This is the self-concept of the company and includes the vision, mission, and corporate values. If these factors of normative management

Fig. 1.9 A systems thinking view of a company, with onion-like layers of elements (Glasl & Kalcher, 2014).

are functional, they also give clear guidance to the formulation of the strategy and the conduct of business activities as well as the overarching reason for the existence of the company in general. Identity is a long-term concept and should remain stable. As the values have a special significance in the context of a social enterprise, we will explain in depth, what we mean by functional corporate values and how they become effective in the setting up of a company and in the interaction with various stakeholders.

1.4.1.2 Strategy

The core of strategic management is a plan for the future of the company that ensures the sustainability and competitiveness of the organization. A strategy is usually a mid-term concept and commonly comprises a thought-through part and an unplanned, emergent part as entrepreneurial activities are carried out in response to perceived internal and external risks and opportunities. The interaction with stakeholders on various levels should be integrated and the corporate values reflected in the formulation of the strategy and the definition of strategic initiatives. This is true in general but is of particular significance in the case of a social enterprise.

1.4.1.3 Organizational structure

Depending on the size, an organization becomes subdivided into smaller units as it grows. This can be done on the level of the whole organization, like in a franchise system or the setting up of new factories, or it can be done by defining new divisions or units. Concepts with a more dynamic approach to organizational structure are becoming more and more popular, like sociocratic structure or holacracy. The subdivision into smaller units as the number of people becomes higher is a consequence of our natural heritage since the human brain can only deal with a limited number of social interactions and at varying levels of complexity (Dunbar, 2016). As the typical pyramidal structure of companies implies a distribution of power according to hierarchical level, it could be essential to consider alternatives right at the start of a social enterprise to counteract the unintended consequence. Although the entrepreneurial drive and vision are required for a social enterprise just like with any conventional enterprises, the implications of setting up a middle management might be unnecessary and undesired. Here, it becomes apparent how a streamlining of functional corporate values with entrepreneurial thinking is mandatory to minimize friction.

Groups and leadership style

This element is the subject of a wide range of discussions in the field of organizational research. On the one hand, humans seem to be more productive in groups and also have the tendency to form a more or less clear hierarchy with a leader when acting in groups. On the other hand, social loafing and groupthink can easily occur and one of the commonly stated reasons for quitting is bosses and their leadership style. Again, it is recommended that before setting up a social enterprise, the way groups are set up should be aligned with entrepreneurial intentions, business requirements, and corporate values. The leadership style, in particular, will often be scrutinized as the "correct" or "respectful" conduct of superiors should be the outcome of a value-based integrity that spans internal and external stakeholders of the social enterprise.

Roles and functions

One of the main reasons to set up a company is to increase the cost-efficiency of production. Subsequently, as the company grows, there is usually a need to diversify the roles and functions of employees. There are two general approaches to achieve this: define positions on business needs and recruit people with appropriate skills and proven performance to fill these positions, or recruit people based on general, cultural, and value fit and create positions for them to develop their potential. The reality is a mixture of these two approaches. For a social enterprise, there seems to be a tendency to favor the latter approach, which may cause obstacles in corporate development in the long run. The risk with the former approach, however, is that well-performing employees may not agree with or act by the corporate values of the social enterprise, which can ruin a corporate culture. In addition, work performance can be culture-dependent and difficult to evaluate across different communities.

Processes

The identity of a company should explain what it does and possibly why. The processes determine the how. Again, there are two general approaches to processes: define and document a process and all its steps and tell employees to follow these or focus on the results and risks involved and let people otherwise do what they feel is necessary or best. And as with the element of Roles and Function, the reality is most often a mixture of these two extremes, however, with shifting ratios in different companies or units. In general, the more intense and complex interactions with other people are, the less strict behavior can be prescribed for a process. For a social enterprise, this entails that if the purpose of the company is directed toward or involves a considerable amount of human interactions, processes should be defined accordingly, providing enough leeway for individual adaptations for employees in pursuing business activities.

Infrastructure and materials

This element encompasses everything physical or virtual required to do the intended business activities. Examples are the production plant, machines, computers, software, or the coffee supply for the staff kitchen. Obviously, a social enterprise should strive for a selection of infrastructure and materials in line with their stated corporate values and in consideration of key stakeholders, e.g. carpenters, IT service providers, or local coffee roasters.

When it comes to corporate development, it is important to realize the interdependencies among the various elements. As one element changes, it triggers further changes, in the same element and other elements, sometimes with a considerable lag. Over time, a development path emerges, specific and unique to an organization. This is why it is so important to consider corporate development straight from the start, as otherwise, the emerging developmental path might be deviating from its desired path.

Although individual traits and features of an organization form spontaneously, in a complex and thus unplannable fashion, general boundary conditions can often be set, e.g., non-profit, non-toxic production, local sourcing. A comprehensive understanding of the coupled elements and factors of the system "social enterprise" is tantamount to ensure sustainability. As a first proxy, any change in one of the elements will cause subsequent changes in all other elements. There might, however, not be enough "difference to make difference" (Bateson, 1970); that is, the changes are sometimes too small to be relevant. At the same time, it is mandatory to examine all elements to test for unintended consequences and making changes in line with the developmental path of the company as well as in line with its functional corporate values. Then, and only then, the preconditions for a social enterprise to perform on the one hand necessary adaptations over the lifecycle of the company in a consistent and integer way, and on the other hand have a solid foundation for transparent and meaningful interactions with the corporation's stakeholders.

1.4.2 Stakeholder management

Generally speaking, an organization does well, if it does something right where it counts. This spot where it counts is for the majority of corporations their product or service. As long as the customers are happy with the properties of the product, they are able to afford it and get it when they need it, they can be expected to be pleased. This traditional, exclusively product-focus perspective, however, had to be replaced on a broader scale by a slightly broader view when customers of the US automobile industry in the early 20th century began to refuse to buy their product, knowing that workers in their own community or township were harmed or even killed during car production. Getting a good product was suddenly not enough. The product had to fulfill other customer expectations as well, e.g., not killing people of their customers' communities during the production process. With time, in a process of social change, expectations expanded and branched out, some becoming mainstream, some being held in subcultures or by individuals only.

Managing such a diverse and to a high extent unknown entity as "our stakeholders" can seem to be an overwhelming task, especially for corporations explicitly focusing on factors beyond their product's functionality. They in particular depend on their stakeholders to recognize their corporate value issues, goals, activities, and norms to fully appreciate and value their corporate existence beyond their product or service provided. Self-aware stakeholders, on the other hand, equally depend on a corporation to declare its corporate values in a structured, transparent, and functional way in order to realize whether the corporation deserves their license to operate or not. It is this codependence of socially oriented corporations and self-aware stakeholders that turns corporate values and their communication into a crucial tool for stakeholder management.

By transparently rooting each and every corporate activity in a corporate value goal, which in turn can be verified to support a corporate issue in a target-oriented manner, a corporation can create a corporate value-oriented culture not only for their internal stakeholders (Schein, 2010; Strautmanis, 2008) but also toward their external ones. In *Cofibo's* case, reaching and informing stakeholders constitute the major part of stakeholder management, as appealing to them and getting a larger market share by remodeling the corporation according to the market's needs are not an option compatible with the vision the corporation is pursuing. The corporation's well-communicated perspective on its role in corporate sustainability and its concept of "doing good" constitutes a blueprint for corporate actions. Internal and external stakeholders can use this blueprint either to provide feedback on it or to validate *Cofibo* is indeed implementing its vision in a way compatible to their individual values, far beyond the particle boards manufactured.

1.4.3 Approaching operations management with a systems thinking mindset

Setting up business operations is a complex affair. It involves numerous stakeholders, production factors, and influencing parameters along its whole value chain. An established way to deal with complexity is the modeling method of system dynamics. Here, we would like to highlight the capabilities and usefulness of this approach without going into details. As many concepts in management are intangible, they require an approximation of their "real" structure. This is undertaken by the formulation of a so-called structural hypothesis, that is, how the elements of the system interact dynamically to constitute a structure. As a conceptual visualization (Fig. 1.6); however, these system dynamic models and the underlying causal loop diagram (CLD) are well-suited to base a discussion on the interdependencies and lagging responses of the system to changes in its overall structural hypothesis or its behavior over time (Sterman, 2010).

Starting from a simple model of the moment production starts, the key elements of this system can be detailed. The system is organized in stock and flows. The stock, in this case, is the actual inventory as well as the work in progress (WIP) inventory of cocoboards. The assumption is that at production start, there is no shortage of material, labor, processing capabilities, or other equipment. Another assumption is a balanced equilibrium of the system; i.e., the stocks are unchanging as the inflow and outflow are equal. Although generic, this model can be easily adapted to a more specific scenario. It allows entrepreneurs and other decision-makers to consider relevant factors in the system of their organization, identify leverage points, and derive policy recommendations (Fig. 1.10).

Factor	Note/Effect
Work in Progress (WIP) Inventory	Stock of half-finished products. Has an inhibiting effect on the Adjustment for WIP
Inventory	Stock of finished products. This factor has an inhibiting effect on the Adjustment for Inventory
Inventory Control	Balancing feedback loop
WIP Control	Balancing feedback loop
Production Start Rate	Rate of flow from an external source to the stock of the WIP Inventory
Production Rate	Rate of flow from the WIP Inventory to the Inventory
Desired Production Start Rate	Increasing effect on the Production Start Rate
Adjustment for WIP	Increasing effect on the Desired Production Start Rate
WIP Adjustment Time	Inhibiting effect on the Adjustment for WIP
Desired WIP	Increasing effect on the Adjustment for WIP
Desired Production	Increasing effect on the Desired Production Start Rate and also on the Desired WIP
Expected Order Rate	External factor that drives the Desired Production and the Desired Inventory
Manufacturing Cycle Time	Increasing effect on the Desired WIP and an inhibiting effect on the Production Rate
Adjustment for Inventory	Increasing effect on the Desired Production
Inventory Adjustment Time	Inhibiting effect on the Adjustment for Inventory
Desired Inventory	Increasing effect on the Adjustment for Inventory
Desired Inventory Coverage	Increasing effect on the Desired Inventory
Safety Stock Coverage	Increasing effect on the Desired Inventory Coverage
Minimum Order Processing Time	Increasing effect on the Desired Inventory Coverage

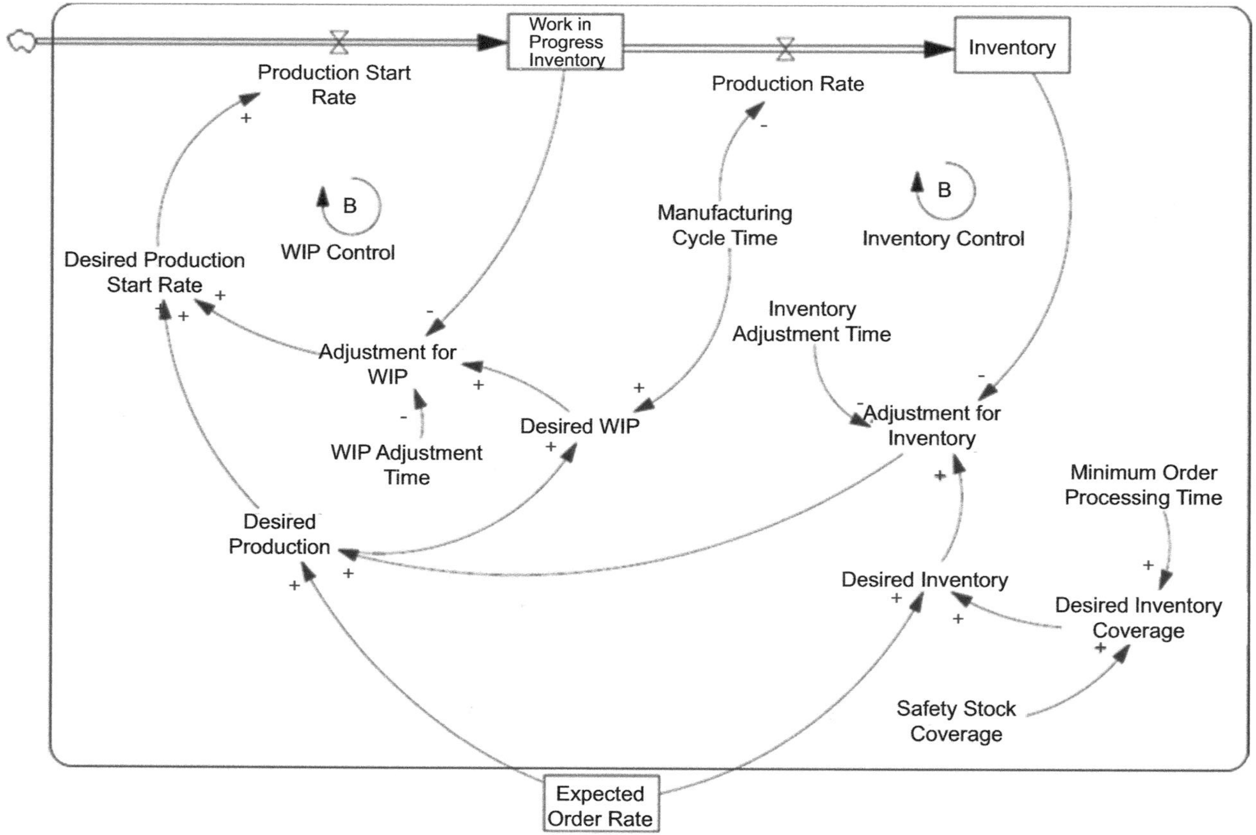

Fig. 1.10 A system dynamics model of the structure of *production starts. Adapted from Sterman, J. (2010). Business dynamics: Systems thinking and modeling for a complex world. Tata McGraw-Hill.*

With this simple model, it is possible to determine the necessary production rate and inventory levels to accommodate current market conditions and customer demand. Similar to the more fundamental systemic model of the whole organization, this model also facilitates taking all relevant factors into account. Additionally, this specific model allows testing scenarios of various market developments and their temporal consequences in production. Finally, this specific CLD of a system dynamics model can be used to run simulations on appropriate software and quantify variables. Subsequently, this system dynamics model is an excellent tool to apply a similar mindset introduced for corporate development to the more hands-on operations management. This is vital to ensure the competitiveness of a social enterprise right from the start. At the same time, this model allows a testing if any changes to production might infringe on the defined functional corporate values and when and if more intense stakeholder management might be necessary, e.g., if smallholder farms cannot sell their coconut husks due to insufficient inventory capacities for raw materials. Thus, the system dynamics model can be a first step to set up a process controlling with metrics and indicators relevant to a social enterprise.

1.4.4 Navigating the business context

Cofibo is designed as a climate-smart and socially and environmentally responsible value chain innovation. The expected growth path is a sound operational management and corporate development based on functional corporate values. Sustainability assessment of setting up a social franchise of a new small-scale industry in a developing country to make the coconut production system more resilient should then be possible. There are three main aspects addressed in the scientific research of *Cofibo* value chain:

(a) contexts (particular to emerging and developing economies);
(b) objectives (economic, environmental, and social responsibility); and.
(c) stakeholders (producers, consumers, shareholders, for-profit/nonprofit/social enterprises, governments, and nongovernmental organizations).

A current literature review on operational management revealed a lack of research in these areas (Lee & Tang, 2017). The sustainability assessment integrates economic, ecological, and social aspects over time. Together, the research will provide a scientific base to evaluate options for making coconut farming more resilient in the face of climate change and its side effects and providing valuable data on social innovations in an emerging market.

A major problem facing the coconut industry is the high incidence of poverty among coconut farm families (Castillo & Quebral, 1996).

The productivity of coconut farming is low in the Philippines. On average, one hectare of coconut yields about a ton of produce, compared to four tons of yield per hectare of cultivated oil palms. Unfortunately, the typical smallholdings are mainly committed to a monocrop of coconut, which occupies the land for about a century. Under such a monocrop system, the majority of the coconut holdings do not generate adequate income and employment for the dependent families. From the land utilization point of view, a pure stand of coconut utilizes only 22% of the area at a spacing of 7.5×7.5 m leaving nearly 78% of the area not effectively utilized (Magat, 1990). Agricultural studies also emphasize the importance of enhancing plant diversity and complexity in farming systems to reduce vulnerability to extreme climatic events (Altieri, Funes-Monzote, & Petersen, 2012). *Cofibo* and complementing projects in the program, therefore, aim at assessing potentials of mixed cropping with tannin-yielding plants (Suharto, 1998). The aim is to improve the livelihood locally by developing a stable value chain for the production of cocoboard technology applications in the Philippines. Communities are trained and introduced to the *Cofibo* franchise, which provides a blueprint to set up a production unit and supply chain for cocoboards to make their farming more resilient in the face of increasing economic pressure but also climate changes. As the pilot production is set up and alternative applications are developed, the potential market is expanded and made accessible. The project uses an integrated approach to minimize the risks of innovation development and market entry and maximize long-term benefits for local communities and the environment. Set up as a social franchise with governance through a social enterprise, *Cofibo* aims to help the poor, who are particularly vulnerable to climate change, to break out of the poverty cycle, and has the backing of relevant stakeholders, generating a win-win situation for coconut farmers, craft people, and everyone involved in the value chain.

The commercial impact is based upon the existing market for particle boards. Cocoboards offer the opportunity to take over a significant part of this market. Moreover, with other applications derived from the basic cocoboard technology, additional value chains can be established, e.g., by integrating long fibers from manila hemp (abacá) (Bajpai, Singh, & Madaan, 2014; Takagi & Hagiwara, 2015). Organized as a social enterprise, the value of this industry is captured locally in the Philippines. Furthermore, as local stakeholders are trained and enabled, further diversification of farming practices can be expected, not only contributing to resilience but also generating new options for income, either through increased production or diversifying their portfolio of farmed species, e.g., growing tannin-yielding crops, or through partaking in other parts of the new value chain.

1.5 Conclusion

This article highlighted some of the main aspects in setting up a social enterprise. Driven by a desire to organize business activities more in line with an alternative to the conventional linear model of economy, the example business of *Cofibo* is described as a company with a value proposition of a low-end innovation, geared to accommodate requirements of a more circular economy operating on a transparent base of corporate values. Its value proposition goes beyond manufacturing a product in a resource-effective and environmentally friendly manner, but also provides added value to their associated communities. The main product of *Cofibo* is an all-natural coconut-based particle board as interior, non-loadbearing building material made from coconut husks. A corporate development model based on systems thinking was then used to detail vital elements of furthering a social enterprise. A particular effort was made to showcase the importance of functional corporate values. Their links to strategy, operations, and stakeholder management were presented and analyzed. One of the most prominent features of a circular economy is the design of value chains in circles. This is reminiscent of the typical mindset of systems thinking and system dynamics. Subsequently, a generic system dynamics model to provide a structural hypothesis of how production could start was presented. Overall, the approach outlined here can be indicative for setting up a social enterprise, with a well-defined relation to sustainability and circular economy. More research is necessary, however, to detail the decision-points for social entrepreneurs in this process and provide appropriate indicators as controlling for the early-stage corporate development. Thus, we hope to contribute to a shift away from the traditional linear "take, make, dispose" product-centered way of doing business to a more sustainable, circular economic system, as well as providing a blueprint to budding social entrepreneurs for starting their enterprises. Furthermore, we strive to propagate a start-up culture

- considering the aspects of their societal "license to operate" and the population of stakeholders they might be addressing with it
- considering the requirements and obstacles in the selected business context
- considering the potential of circular business models and its implementation
- considering how the selected corporate values narrow down the boundary conditions for corporate development

Overall, the example of the social enterprise *Cofibo* illustrates how the combination of applied science and entrepreneurial thinking and acting might contribute to a larger societal change and a more sustainable development.

References

Altieri, M. A., Funes-Monzote, F. R., & Petersen, P. (2012). Agroecologically efficient agricultural systems for smallholder farmers: Contributions to food sovereignty. *Agronomy for Sustainable Development*, *32*, 1–13.

Austin, J., Stevenson, H., & Wei-Skillern, J. (2012). Social and commercial entrepreneurship: Same, different, or both? *Revista de Administração*, *47*, 370–384.

Bajpai, P. K., Singh, I., & Madaan, J. (2014). Development and characterization of PLA-based green composites: A review. *Journal of Thermoplastic Composite Materials*, *27*, 52–81.

Bateson, G. (1970). Form, substance and difference. *General Semantics Bulletin*, *37*, 5–13.

Blewitt, J. (2008). In J. Blewitt (Ed.), *Community, empowerment and sustainable development* (p. 192). Green books.

Castillo, E., & Quebral, N. (1996). Modernizing the coconut industry: Problems, issues, and prospects. In *Paper submitted to Congressional Commission on Agricultural Modernization (AGRICOM)*. Quezon City, Phillipines: Agricom.

Christen, M. (2010). A theory of the good for a conception of sustainability. In *The sixteenth annual international sustainable development research conference conference proceedings, Hong Kong*.

Dacin, P. A., Dacin, M. T., & Matear, M. (2010). Social entrepreneurship: Why we don't need a new theory and how we move forward from here. *Academy of Management Perspectives*, *24*, 37–57.

Dees, J. G. (2018). The meaning of social entrepreneurship 1, 2. In *Case studies in social entrepreneurship and sustainability* (pp. 22–30).

Dunbar, R. (2016). 11 Robin dunbar on dunbar numbers. In D. Edmonds, & N. Warburton (Eds.), *Big ideas in social science* (pp. 102–109). SAGE Publications Ltd. 1 Oliver's Yard, 55 City Road London EC1Y 1SP.

Ellen MacArthur Foundation. (2017). *What is a circular economy?*. Ellen MacArthur Foundation. [Internet]. [cited 23 Apr 2019]. Available https://www.ellenmacarthurfoundation.org/circular-economy/concept.

Frecè, J. T. (2019). *Corporate values—A socio-functionalist approach exemplified along corporate sustainability values*. University of Basel.

Frecè, J. T., & Harder, D. L. (2018). Organisations beyond Brundtland: A definition of corporate sustainability based on corporate values. *Journal of Sustainable Development*, *11*, 184–193.

Freeman, R. E. (1984). *Strategic management: A stakeholder approach*. Boston: Pitman.

Friedman, M. (1970). *The social responsibility of business is to increase its profits*. The New York Times Magazine.

Fritsch, M. (2019). *Entrepreneurship*. https://doi.org/10.1007/978-3-662-57984-8.

Gassmann, O., & Sutter, P. (2013). *Praxiswissen Innovations management: Von der Idee zum Markterfolg*. Carl Hanser Verlag GmbH Co KG.

Geissdoerfer, M., Savaget, P., Bocken, N. M. P., & Hultink, E. J. (2017). The Circular Economy—A new sustainability paradigm? *Journal of Cleaner Production*, *143*, 757–768.

Glasl, F., & Kalcher, T. (2014). *Professionelle Prozessberatung: Das Trigon-Modell der sieben OE-Basisprozesse*. Verlag Freies Geistesleben.

Goodland, R., & Ledec, G. (1987). Neoclassical economics and principles of sustainable development. *Ecological Modelling*, *38*, 19–46.

Graedel, T. E., & Allenby, B. R. (2003). *Industrial ecology*. Prentice Hall.
Gunderson, L. H. (2000). Ecological resilience—In theory and application. *Annual Review of Ecology and Systematics, 31*, 425–439.
Harder, D. L., & Tokarski, K. O. (2018). The power to change a social system. In C. Machado, & J. P. Davim (Eds.), *Organizational behaviour and human resource management: A guide to a specialized MBA course* (pp. 49–72). Cham: Springer International Publishing.
Horlings, L. G. (2015). The inner dimension of sustainability: Personal and cultural values. *Current Opinion in Environmental Sustainability, 14*, 163–169.
Huybrechts, B., & Nicholls, A. (2012). Social entrepreneurship: Definitions, drivers and challenges. In C. K. Volkmann, K. O. Tokarski, & K. Ernst (Eds.), *Social entrepreneurship and social business: An introduction and discussion with case studies* (pp. 31–48). Wiesbaden: Gabler Verlag.
Intergovernmental Panel on Climate Change. (2014). *Climate change 2014—Impacts, adaptation and vulnerability: Regional aspects*. Cambridge University Press.
International Tropical Timber Organization. (2004). *Introduction of a village industry in the community around an industrial forest plantation in Indonesia*. Technical report Bogor, Indonesia: SEAMEO BIOTROP. Report No.: 58/99 Rev. 1(I) https://www.itto.int/files/itto_project_db_input/2331/Competition/PD-58-99-R1-I-Completion-Report.pdf.
Kalmykova, Y., Sadagopan, M., & Rosado, L. (2018). Circular economy—From review of theories and practices to development of implementation tools. *Resources, Conservation and Recycling, 135*, 190–201. https://doi.org/10.1016/j.resconrec.2017.10.034.
Lee, H. L., & Tang, C. S. (2017). Socially and environmentally responsible value chain innovations: New operations management research opportunities. *Management Science, 64*, 983–996.
Magat, S. S. (1990). Growing conditions and growth habits of coconut in relation to coconut-based farming systems. Coconut based farming systems. In S. de Silva (Ed.), *XXVII COCOTECH Meeting* (pp. 25–29). Asian & Pacific Coconut Community.
McDonough, W., & Braungart, M. (2010). *Cradle to cradle: Remaking the way we make things*. Farrar, Straus and Giroux.
McKelvey, B. (2002). Managing coevolutionary dynamics. In *18th EGOS conference* (pp. 4–6). Barcelona, Spain: billmckelvey.org.
Nawir, A. A., & Santoso, L. (2005). Mutually beneficial company-community partnerships in plantation development: Emerging lessons from Indonesia. *International Forestry Review*, 7, 177–192.
Osterwalder, A., & Pigneur, Y. (2010). *Business model generation: A handbook for visionaries, game changers, and challengers*. John Wiley & Sons.
Paswan, A. K., & Wittmann, C. M. (2009). Knowledge management and franchise systems. *Industrial Marketing Management, 38*, 173–180.
Pauli, G. A. (2010). *The blue economy: 10 years, 100 innovations, 100 million jobs*. Paradigm Publications.
Philippine Coconut Authority (PCA). (2013). *A briefing guide on the subject: Status, "coconut industry production; zones, growing; technologies, efficient farming"*. Republic of the Philippines: Department of Agriculture. Available http://www.pca.da.gov.ph/coconutrde/images/gen8.pdf.
Reinhardt, R., Gurtner, S., Hoskins, J. D., & Griffin, A. (2017). The high-end bias—An irrational preference of decision makers for high-end over low-end innovations. *Academy of Management Proceedings, 2017*(1), 13528.
Richard, G. A., Frazer, L., Weaven, S. K., & Dant, R. P. (2016). Building franchisee trust in their franchisor: Insights from the franchise sector. *Qualitative Market Research: An International Journal, 19*, 65–83.
Schein, E. H. (2010). *Organizational culture and leadership* (4th ed., p. 436 S). San Francisco: Jossey-Bass.

Segelitz-Karsten, A., Reinhardt, R., Hietschold, N., Gurtner, S., & Walsh, G. (2019). Attitude towards low-end and high-end innovation. *Academy of Management Proceedings*, *1*, 18945. Academy of Management Briarcliff Manor, NY 10510.

Simon, F. B. (2004). *The organisation of self-organisation: Foundations of systemic management.* Carl Auer.

Siregar, U. J., Rachmi, A., Massijaya, M.Y., Ishibashi, N., & Ando, K. (2007). Economic analysis of sengon *(Paraserianthes falcataria)* community forest plantation, a fast growing species in East Java, Indonesia. *Forest Policy and Economics*, *9*(7), 822–829.

Smith, P. A., & Sharicz, C. (2011). The shift needed for sustainability. *Learning Organization*, *18*(1), 73–86.

Stahel, W. R. (2016). The circular economy. *Nature*, *531*, 435–438.

Sterman, J. (2010). *Business dynamics: Systems thinking and modeling for a complex world.* Tata McGraw-Hill.

Strautmanis, J. (2008). Employees' values orientation in the context of corporate social responsibility. *Baltic Journal of Management*, *3*, 346–358.

Suharto, J. C. (1998). Potentials for increasing farmers' income and enhancing competitiveness of the coconut industry through alternative uses. In P. A. Batugal, V. Ramanatha Rao, & C. Bong (Eds.), *Promoting multi-purpose uses and competitiveness of the coconut* (pp. 26–29). Serdang: IPGRI-APO.

Takagi, H., & Hagiwara, Y. (2015). Fracture behaviour of natural fibre reinforced composites. In *Vol. 87. WIT transactions on state-of-the-art in science and engineering* (pp. 195–204). WIT Press.

Tsay, A. A., Nahmias, S., & Agrawal, N. (1999). Modeling supply chain contracts: A review. In S. Tayur, R. Ganeshan, & M. Magazine (Eds.), *Quantitative models for supply chain management* (pp. 299–336). Boston, MA: Springer US.

Tukker, A., Bulavskaya, T., Giljum, S., de Koning, A., Lutter, S., Simas, M., et al. (2016). Environmental and resource footprints in a global context: Europe's structural deficit in resource endowments. *Global Environmental Change*, *40*, 171–181.

UN Environment: International Resource Panel. (2017). *Decoupling natural resource use and environmental impacts from economic growth.* [Internet]. 3 Feb. [cited 14 Jan 2020]. Available: https://www.resourcepanel.org/reports/decoupling-natural-resource-use-and-environmental-impacts-economic-growth.

Webley, S. (1999). Sources of corporate values. *Long Range Planning*, *32*, 173–178.

Yap, E. G. O. (2013). Socio-cultural factors affecting the market behavior of coconut farmers in Central Philippines. *IAMURE International Journal of Business and Management*, *6*(1), 31–44.

CHAPTER 2

Blockchain technology and socially sustainable supply chains—A valuation perspective

Aref Aghaei Hervani[a], Santosh Nandi[b], Joseph Sarkis[c], and Marilyn M. Helms[d]
[a]Lander University, Greenwood, SC, United States
[b]Walker College of Business, Appalachian State University, Boone, NC, United States
[c]School of Business, Worcester Polytechnic University, Worcester, MA, United States
[d]C. Lamar and Ann Wright School of Business, Dalton State College, Dalton, GA, United States

2.1 Introduction

Corporations and consumers have grown more concern about social and environmental sustainability (Groening, Sarkis, & Zhu, 2018). Organizations are faced with incorporating environmental or social sustainability decisions both globally as well as along their supply chains, and many times with the latest emergent technologies (Sarkis & Zhu, 2018). These practices are a culmination of various social, cultural, political, technological, and economic forces causing organizations to rethink how business is conducted and assessed. With the emergence of blockchain technology, social sustainability can more likely effectively address through capabilities that include decentralized and immutable data, reliable data, transparency, traceability, and smart contracts, and incentivization is also possible.

A blockchain is—in the simplest of terms—a time-stamped series of immutable records of data managed by a cluster of computers not owned by any single entity. Each of these blocks of data is secured and bound to each other using cryptographic principles—the chain. The technology, already popularly used with Bitcoin and other cybercurrencies, has industry-disrupting capabilities for social sustainability across supply chains. The blockchain network has no central authority. Since it is a shared and immutable ledger, the information in it is open for participants in the supply chain to access and see. Anything that is built on the blockchain is meant to be transparent tool involved; it holds members accountable for their actions. This is a crucial tenet of traceability necessary for global supply chain management.

Green Production Engineering and Management
https://doi.org/10.1016/B978-0-12-821238-7.00007-5

With infrastructure costs, a blockchain typically carries no or minimal transaction cost. The blockchain helps pass essential information from A to B in a fully automated and relatively secure manner. One party to a transaction initiates the process by creating a block. Verification of the block of information can be completed by a set of participants, depending on whether it is a permissioned or public blockchain. A permissioned blockchain is a private—membership only—blockchain; a public blockchain is open to anyone (Kouhizadeh & Sarkis, 2018). The verified block is added to a chain, which is stored in multiple locations, creating not just a unique record, but a unique record with a unique history. Falsifying a single record would mean falsifying the entire chain in millions of instances, which is virtually impossible—thus the security and reliability of data. Bitcoin uses this model for monetary transactions as a public blockchain. But blockchain technology can be deployed in many other ways as this chapter will explore and link social sustainability, global supply chain management (GSCM) and performance and valuation.

This chapter provides both a contribution to theory and practice. We provide examples of blockchain technology's advantages for ensuring and supporting social sustainability through valuation approaches, within organizations and across supply chains. Researchers and practitioners are able to use our proposed framework to guide them in how blockchain can support valuation of social sustainability in supply chains. Researchers are provided with some possible research directions on this complexly evolving topic.

2.2 Blockchain technology (BCT)—Background and elements

Blockchain technology (BCT) is viewed as a disruptive technological innovation, especially in the supply chain domain due to its potential to enhance the efforts of supply chain digitalization and Industry 4.0 standardization (de Sousa Jabbour, Jabbour, Foropon, & Godinho Filho, 2018; Saberi, Kouhizadeh, Sarkis, & Shen, 2019; Wang, Han, & Beynon-Davies, 2019). With several definitions around, the characterization of blockchain technology is no different from "the blind men and the elephant" analogy—where every blind man has a different perspective on the elephant (Saberi et al., 2019). Through our survey of the literature, we identified four distinct elements that reflect how blockchain technology can operationalize future supply chains. The four elements are: BCT as a decentralized consensus mechanism, BCT as an information-sharing mechanism, BCT as an incentivizing mechanism, and BCT as a negotiating mechanism, as described below.

2.2.1 BCT as a decentralized consensus mechanism

Rooted in Satoshi Nakamoto's (2009) seminal paper, the concept of blockchain is based on two primary attributes, namely, a *distributed ledger system* and *cryptographic tools* (Nakamoto, 2009). The *distributed ledger system* provides the verification process for transactions using a predefined consensus method among the participating entities, thus bypassing the third-party auditing requirement (Yermack, 2017). *Cryptographic tools* provide a rigorous verification process to transform each transaction into irreversible hexadecimal token data. This irreversible process ensures that the data are immutable and secure. No individual entity can alter the verified and recorded information once it becomes a part of the distributed ledger system (Yermack, 2017). Combining the distributed ledger and cryptographic tools, the transactions of a blockchain technology-based system become immutable, secure, and tokenized. Thus, *cryptographic mechanisms* are formed and are self-organized as decentralized consensus rule-based *distributed ledgers* for the participating networks of entities across geographies.

2.2.2 BCT as an information-sharing mechanism

The information-sharing element of blockchain technology derives from how distributed ledgers of information are structured to be transparently shared among the participating entities based on openness of the platform to act as permission less (public) or with permissioned (private) or a combination of both (Ølnes, Ubacht, & Janssen, 2017). As transactions occur, they are cryptographically time-stamped, thus enabling participating entities to track each record, alongside its creating body, and track the secured information according to the blockchain information dimensions (Swan, 2015). The timestamp improves the authenticity of recorded information about the product or service and its properties because it can verifiably describe what, when, where, why, by whom, and for whom a transaction occurred, say, in a typical blockchain-based supply chain system (Kelley, 2018). For example, when a firm that sells processed meat implements a blockchain-based ledger system across its supply chain, its customers can access the knowledge about the processed meat product. Data could include information about product origin, production, modifications, and ownership, which in turn enables customers to be confident in their purchasing decisions.

A blockchain technology-based system can be coupled with other secured information-capturing devices, including radio frequency identification (RFID), the Internet of Things (IoT), and global position sensors (GPS). The coupling allows the capture and flow of information with accuracy and in real-time (Kshetri, 2017). These integrations designed to capture verifiable, immutable, and reliable data address both transparency and traceability issues in complex supply chain networks. As an outcome of such a blockchain-based information sharing mechanism, the need for third-party transaction verification is eliminated, thus making the information self-sustainable (Francisco & Swanson, 2018).

2.2.3 BCT as negotiating mechanism

The negotiating BCT mechanism element is derived from blockchain's ability to mirror real-world negotiations. This mirroring occurs by cryptographically tokenizing the codes of agreements between parties, monitoring conditions of fulfillment, and self-executing the performance functions. Performance functions may include work orders, payment, and similar documents upon the fulfillment of obligations according to the binding agreements between two or more parties (Macrinici, Cartofeanu, & Gao, 2018). The concept of a *smart contract* is grounded upon this process of digitally executing a real-world contract in a decentralized blockchain network that stores rules and policies for negotiating contract terms, can verify the contract automatically, and can execute them upon agreed conditions by the system of participating entities after reaching consensus on the outcomes of the agreement (Saberi, Kouhizadeh, & Sarkis, 2018). Two key benefits of smart contracts include replacing the need for a traditional legal centralized entity with network consensus and increasing efficiency and reducing transaction cost by applying self-executable contract functions.

2.2.4 BCT as incentivizing mechanism

The incentivization element of blockchain technology requires firms to structure incentives for their participating entities. Depending on the scope of blockchain-based applications, firms will implement incentive systems. Such systems will encourage people, partners, and stakeholders adopt new products, processes, or outcomes and also validate information records—also referred to as mining (Kouhizadeh, Sarkis, & Zhu, 2019). Such incentive systems can enhance the performance of the overall distributed-ledger environment, especially data integrity, product management policies, and data acquisition. Forms of incentives can be financial—such as through

Bitcoin and cryptocurrencies—or non-financial—such as reward tokens and knowledge coins—incentives that can be exchanged into financial form or vice-versa (Chen, 2018).

2.3 Social sustainability elements

Sustainability includes economic, environmental, and social dimensions that comprise the triple bottom line of an organization. Environmental and social responsibility criteria can be measured quantitatively or qualitatively, to judge the overall performance of a company. The general agreement is that implementing and managing economic, social, and environmental efforts of business sustainability should be connected in a balanced and comprehensive way (Svensson & Wagner, 2015). Yet effectively, for business purposes, economic factors typically supersede the social and environmental, with social sustainability remaining the least integrated measure (Schönborn et al., 2019).

While it is crucial to consider social sustainability, it is challenging to address, and may also include environmental factors. Unlike economic dimensions, there are more cultural and intangible characteristics present in the social sustainability dimension, making the assessment more conducive for indirect and intangible evaluations (Varsei, Soosay, Fahimnia, & Sarkis, 2014). Social sustainability for organizations can be supported through the use of tools for valuation and performance. Much of the research, as well as public pressure concerning sustainability, focuses on the effects of business and organizational activity on the physical environment (Schönborn et al., 2019). Still, companies and their practices affect the human and social environment just as often as the physical environment. There are both direct and indirect effects of organizations concerning decisions about people, human health, social systems, and mortality. Yet, social sustainability has received limited coverage by sustainability researchers, especially relating to organizational supply chain (Brandenburg, Govindan, Sarkis, & Seuring, 2014; Mani, Jabbour, & Mani, 2020).

Marnewick and Labuschagne (2005) proposed a framework of social sustainability criteria relevant to projects over their entire life cycle and included: (1) Internal Human Resources, (2) External Populations, (3) Stakeholder Participation, and (4) Macro-Social Performance issues. We further agree that social aspects are rarely considered, while environmental factors are typically considered separately, although they are part of many corporate social responsibility and social sustainability criteria.

Many difficulties do exist in managing social sustainability. There is difficulty in coordinating activities and objectives of internal functions or members of the supply chain, resistance by either employees or supply chain members including suppliers or direct customers. Companies often do not include corporate social responsibility (CSR) criteria in their requirements, so no CSR information is obtained from their suppliers. It is difficult for a firm to check the entire supply chain and identify the source of supplier products from sensitive areas of the world.

Existing models of social sustainability do consider both internal employee stakeholders and external stakeholders and populations or include measures of business success and social performance. These social sustainability factors will be utilized as a starting point in developing a framework for environmental goods valuation integration, blockchain technology, and supply chains that appear in this chapter.

2.4 Market and non-market valuation methods

Environmental and many social goods are not traded in markets, and the valuation of environmental and social assets requires the use of a technique that enables a researcher to identify the link that exists between market goods and social goods and then utilize this link in evaluating the welfare changes associated with changes in social and environmental quality. The most popular way to obtain the value of public—social—goods has been elicited by survey methods enabling researchers to obtain consumer maximum *willingness to pay* for a public good. Willingness to pay is the maximum amount of income a person would pay in exchange for an improvement in circumstances and conversely *willingness to accept* is the minimum income a person would accept for a decline in circumstances. The theoretical framework upon which environmental valuation is founded is based on consumer welfare theory. Consumer surplus was first defined as the residual difference between the value of a consumption bundle to the consumer and the amount which that consumer actually paid for it (Marshall, 1930).

There are two approaches to estimate the use and the non-use components of the total economic value of social and environmental goods: market-based and non-market-based valuation techniques. These two approaches can integrate intangible and tangible valuations for environmental and social benefits. Use and non-use values that individuals place on environmental resources are measured through techniques that use either physical or behavioral linkages (Mitchell & Carson, 1989).

The *physical linkage method* is a "dose-response model" and enables an analyst to obtain changed use and welfare by establishing relationships with physical causes. By examining the relationship between levels of pollution and the physical damage effect, for example, analysts determine a physical damage function, assign prices, and estimate the economic value of the damage that accrues.

The *behavioral linkage method* focuses on behavior and welfare changes. Behavioral changes may either be revealed or stated; the changes may be directly or indirectly linked to the environmental resources being valued. The non-market valuation techniques use the behavioral linkage approach to measure environmental benefits, while the market valuation techniques are primarily viewed from the physical linkage approach. Table 2.1 provides a summary of market and non-market valuation techniques for measuring the environmental benefits important to this research—these techniques can play important roles for social goods valuation and socio-environmental sustainability.

2.5 Blockchain-based socio-environmental valuation framework

As shown in Fig. 2.1 and Table 2.2, a framework that starts to link columns representing blockchain technology elements—decentralized consensus, information sharing, negotiating, and incentivizing mechanisms—to socio-environmental sustainability dimensions. The rows of the framework represent various socio-environmental sustainability indicators including internal human resources, external populations, stakeholder participation, and macro social issues.

Internal human resources issues include employment stability, employment practices, health and safety, and capacity development. The external population includes factors such as human capital, productive capital, and community capital. Information provisions and stakeholder influence are the two components of stakeholder participation. Finally, macro-social issues include the socio-economic performance and socio-environmental performance measures.

BCT can aid multiple supply chain stages (Kouhizadeh & Sarkis, 2018; Saberi et al., 2019). At different nodes along the supply chain, the information provides valuable insights to supply chain operations from the most basic raw material provider to end-user consumers and even back again through reverse logistics networks. As mentioned, blockchain can provide

Table 2.1 Market and non-market valuation methods.

Behavioral linkage approach: indirect estimation methods	Definition
Revealed preference methods	An approach that is used to identify the underlying preferences, and thus demands of individuals, based on the choices each reveals in their consumption
Hedonic pricing	The method assesses the value of an environmental feature (clean air, clean water, serenity, view) by examining actual markets where the feature contributes to the price of a marketed good
Travel costs	The method (TCM) estimates economic values associated with ecosystems or sites that are used for recreation. It assumes that the value of a site can be deduced from how much people are willing to pay to travel to visit the site
Averting expenditures	It can be utilized to measure actions and expenditures incurred to reduce risks. In some cases, it is possible to engage in actions that reduce risks (e.g., staying indoors in days with high air pollution) but it is not easy to place a monetary value on these actions
Random utility model	The model (RUM) is a model of consumer choice in which the consumer is assumed to have perfect discrimination capability between goods or activities in order to maximize their "utility" (relative attractiveness of competing alternatives)
Stated preference methods	Utilized when the researcher does not have actual data on behavior with regard to a certain environmental good or service. The individuals are provided with hypothetical scenarios, based on plausible outcomes and options, and their choices are used to determine the value of the environmental good or service
Contingent valuation	The CVM or CV usually takes the form of a survey questionnaire, which elicits values for environmental goods and services based on hypothetical situations
Contingent choice method	The method can be used to estimate economic values for virtually any ecosystem or environmental service and can be used to estimate non-use as well as use values
Calibrated and conjoint analysis	Statistical technique used to determine how people value different features that make up an individual good or service; it can be used to determine the values attributed to different dimensions of an environmental resource

Discrete choice	This method uses models of consumer choice in which the good or alternative chosen by the consumer is available only in discrete (integer) units
Referendum method	This is a survey method in which the respondent is asked to respond "yes" or "no" to a hypothetical tradeoff between some amount of environmental good or service and something else of value (typically money)
Paired ranking	The method of paired comparisons is used to measure individual preference orderings of items presented to them as discrete binary choices
Choice experiments	Estimates economic values for virtually any ecosystem or environmental service by asking people to make tradeoffs among sets of ecosystem or environmental services or characteristics. Choice experiments do not directly ask for willingness to pay; this is inferred from tradeoffs that include cost as an attribute
Physical linkages approach: direct estimation methods	
Cost-based method or expenditure analysis	Expenditure Analysis examines indirect expenditures that are tied to environmental resources, which are often left out of many traditional analyses
Productivity method	The Productivity Method is used to estimate the economic value of ecosystem products or services that contribute to the production of commercially marketed goods
Cost of illness	Can be utilized to measure Human Health Effects. Relatively easy to perform but fails to capture the value of the discomfort of being sick
Avoided cost method	The Avoided Cost Method calculates the economic value of benefits that an ecosystem provides that would not exist without the ecosystem in place, and therefore, would represent an added cost to society if this environmental service no longer existed
Damage assessment model	The Damage Assessment Model is much like the Avoided Cost Method. The model uses a damage function to calculate the environmental and social costs of alterations to the natural environment

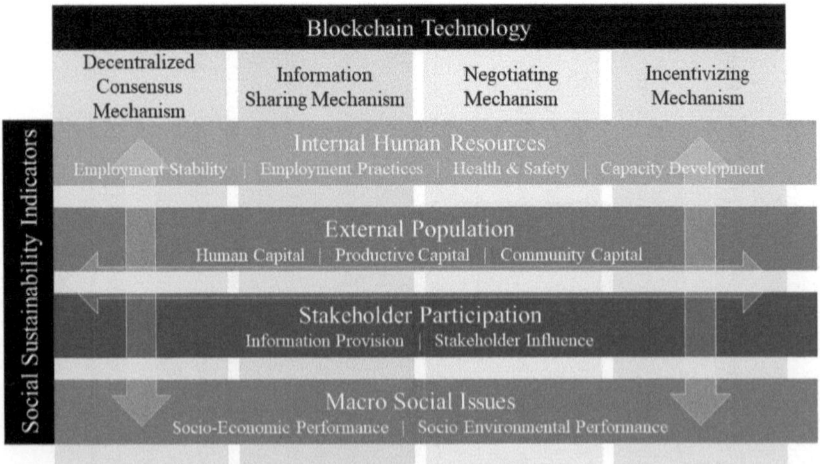

Fig. 2.1 Block chain and social sustainability flows and indicators.

multiple stakeholders throughout the supply chain with reliable, secure, transparent data and information, plus the other characteristics which can help in valuation of activities, resources, materials, products, and processes along the blockchain. Many of these aspects can be used to value socio-environmental sustainability of a supply chain. We now introduce an overview of the non-market valuation methods to help further describe the framework.

2.6 The proposed framework

Table 2.1 includes market and non-market environmental goods valuation techniques for environmental benefit measurement with definitions of each approach; these techniques are applicable to social sustainability dimensions as well. In Table 2.3, we populate Fig. 2.1 and Table 2.2 frameworks with use cases or examples of the relationship among BCT, socio-environmental sustainability, supply chains, and economic valuation methods. We can integrate the market (MVM) and non-market valuation methods (NMVMs) to show how they can be evaluated by or used to evaluate BCT at organizational and supply chain levels(Hervani, Sarkis, & Helms, 2017; Sarkis & Hervani, 2017; Sarkis, Hervani, & Helms, 2013). In other words, we are using market and non-market valuation methods to measure BCTs benefits or how MVMs and NMVMs can be supported using BCTs.

Most organizational performance measurements focus on financial, quantitative, and tangible measures (Burritt & Schaltegger, 2010;

Table 2.2 Social sustainability, supply chain, and block chain technology interrelationships.

	Block chain technology			
Indicators of social sustainability	**Decentralized consensus mechanism**	**Information sharing mechanism**	**Negotiating mechanism**	**Incentivizing mechanism**
Internal human resources				
Employment stability Employment practices Health and safety Capacity development				
External population				
Human capital Productive capital Community capital				
Stakeholder participation				
Information provision Stakeholder influence				
Macro social issues				
Socio-economic performance Socio-environmental performance				

Table 2.3 Indicators of social sustainability and relationships to block chain technology.

Indicators of social sustainability	Blockchain technology			
	Decentralized consensus mechanism	Information sharing mechanism	Negotiating mechanism	Incentivizing mechanism
Internal human resources—*How would employees value its benefits?*				
Employment stability	The value employees have from BCT system's decentralized information sharing mechanism in permissioned/permission-less setting may increase their trust on the new system. The value of this trust along the supply chain may be determined through a *Contingent Valuation Method*. Increasing trust within and between organizations increases employment stability		BCT can provide easy tools that can provide ample opportunities for its employees to operate more effectively and efficiently across the supply chain. Less time in negotiating contracts can lead to greater productivity and value added is measured through the *Productivity Method*	BCT operation can provide employment compensation when employees are out-of-work due to injury or sickness. This support can also occur for a person no matter where they are as long as they have electronic connections. The question here would be "What is the monetary value to employee?" which can be evaluated using a *Contingent Valuation Method*
Employment practices	BCT enables decentralized distribution of common employee practices and can be valued using *Calibrated and Conjoint Analysis* BCT enables employee practices in one location to be shared securely with other locations along the supply chain. We can determine this employee benefits through the *Contingent Valuation Method*		A BCT system—can provide better assurance of human rights and fair work practices both internally and across the organization's supply chain. This characteristic can be valued using a *Contingent Choice Method*	A BCT system can enable employees to securely work from home or outside locations, and thus enables employees not to lose compensation when employees are out-of-work due to injury or sickness. It can help employees see previous effort and work across the supply chain using the *Contingent Valuation Method*

Health and safety	Data verification and information availability allow the organization to address on-going health and safety practices with its BCT system; it can do this across the supply chain. This can lower cost by eliminating further actions to lower the health risk which can be measured using the *Averting Expenditure approach*. Records of incidents can also be collected and storied historically to arrive at values	A BCT operation can enable employees to trace the safety of the workplace and machineries unrestrictedly and determine valuations using the *Avoided Cost Method*	Relates to safety and health outcomes of specific BCT operations involving employees' health and safety adjustments (i.e., Hazardous waste encounters in different departments of a "tire production system"). This can be assessed by applying the *Cost of Illness Method*
Capacity development	BCT operation provides decentralized and reliable data at each of production stage in the supply chain that is protected from getting hacked. The socio-environmental benefits of such capacity related supply chain activities can be captured using the *Contingent Valuation Method*	Through information sharing mechanism that BCT provides, organizations can help improve equity and reduce discrimination at workplace. The assessment of users' willingness to pay for the improved equity can be done using the *Contingent Choice Model* BCT can track supply chain processes to inform users about the state of the product (i.e., reliability and safety). The value of this added incentive can be assessed using the *Avoided Cost Method*	BCT systems enable supply chain members to participate in promoting social and environmental sustainability in its supply chain activities by applying negotiation and incentivization mechanisms. The enhancement in value of the overall capacity development aspects of the focal firm can be measured through the *Productivity Method*

Continued

Table 2.3 Indicators of social sustainability and relationships to block chain technology—cont'd

Indicators of social sustainability	Blockchain technology			
	Decentralized consensus mechanism	**Information sharing mechanism**	**Negotiating mechanism**	**Incentivizing mechanism**
External population—***How would external population value its benefits?***				
Human capital	An organization can leverage on BCT systems that can systematically inform the community about reduction in landfill disposals that helps in improving the quality of drinking water and avoiding environmental and social health damages to the community. The value incurred at the geographic location can be measured in monetary terms using the *Avoided Cost Method*. A BCT system can distribute and inform the on-going human capital gains of sustainability driven activities, such as improvement in air-quality, health of workers. The value of such human capital gains can be assessed using the *Contingent Valuation Method*		The BCT driven career portal of an organization can create a reliable and trusted one-stop source for people in and outside the regions of operations to apply and secure jobs. The socio-environmental value created by the firm by virtue of these job creations at the location can be assessed using the *Productivity Method*	BCT enables members of the manufacturing process of a product to compete to earn incentives for helping in reducing greenhouse gases emissions. The costing method to understand the willingness to pay for such greener products can be measured using the *Averting Expenditure Method*
Productive capital	The BCT-enabled systems of organization can track and distribute regulatory and public service activities, such as stop accidental spillages or leakages that can ultimately cause health and possible ecological damages. The cost to the community that BCT enabled systems can achieve by protecting the ecological systems can be assessed using the *Cost of Illness Method*	When an organization supports on-going community projects through BCT's information sharing requirements, it can help communities to attract more donations from stakeholders, which in turn makes a community more safe and secure. The cost of risk getting averted through such BCT based action can be measured using the *Averting Expenditure Method*	When an organization supports production activities through the BCT system, it encourages its partners to negotiate for greater productivity and efficiency. The estimates of productivity gain in monetary terms can be achieved using the *Contingent Valuation Method*	By adopting BCT, an organization and its supply chain partners can earn economic performance incentives. The value of added performance that it might be willing to transfer to its customers can be measured using the *Contingent Valuation Method*

Community capital	BCT leads to traceability and distribution of information, which helps in achieving greater sustainability results through assurance of adopted human rights, and fair, safe work practices. The value that it brings in the form of community capital gains can be assessed in monetary terms using the *Contingent Valuation Method*		An organization's BCT operation can incentivize communities by allowing them to negotiate with organizations to reduce the flow of waste disposal and divert waste from landfills, reducing contaminated run-offs which in turn lowers risks of being exposed to certain toxins. The quantum of trade off that the community requires to do through more consumption of recycled goods can be assessed using *Choice Experiment*	
Stakeholder participation—*How would stakeholders value its benefits?*				
Information provision	BCT enables supply chain agents to extract data that are decentralized and traceable. Stakeholders can assess the value of availability of the information by using *the Contingent Valuation Method*	BCT operations of an organization can allow stakeholder to share information, thus empowering them to gain greater awareness and participation towards ecological presentation activities in the community at large. The value placed on this activity of the organization can be estimated using the *Contingent Valuation Method*	BCT can engage stakeholders to negotiate about actions required to achieve greater ecosystem benefits and avoid additional costs to society. An *Expenditure Approach* can be used to measure the value derived out of such negotiations Additionally, an *Avoided Cost Method* can be applied to assess the cost of risks through BCT based practices for waste reduction in a societal setting	BCT system can devise incentivization mechanism for stakeholders who actively participate in sustainability driven product stewardship. A *Random Utility Method* can be applied to assess the value derived from such collectiveness in audience participations
Stakeholder influence	BCT of an organization can help stakeholders to record, distribute, trace, and share information about activities focused on maintaining ecological balance. An *Expenditure Approach* can be used to measure the avoid additional costs to the society, in monetary terms. Additionally, a *Contingent Choice Model* can be used to assess the quantum of wastes generated and energy minimized in monetary terms		BCT operations can influence stakeholders support its environmental/societal activities such as tracking government's clean city efforts on decision-making parameters. Using an *Expenditure Approach*, the value of such benefits resulting from societal influences can be estimated in monetary terms	Gathering multiple stakeholder input to understand the expected incentives of cost-savings and risk aversion (post-BCT implementation) can make BCT-based systems more meaningful. The value derived from this activity can be measured in monetary terms using the *Discrete Choice Method*

Continued

Table 2.3 Indicators of social sustainability and relationships to block chain technology—cont'd

<table>
<tr><th rowspan="2">Indicators of social sustainability</th><th colspan="4">Blockchain technology</th></tr>
<tr><th>Decentralized consensus mechanism</th><th>Information sharing mechanism</th><th>Negotiating mechanism</th><th>Incentivizing mechanism</th></tr>
<tr><td colspan="5">Macro social issues—How would society value its benefits?</td></tr>
<tr><td>Socio-economic performance</td><td>BCT capabilities of an organization can enhance the overall corporate and regional trading opportunities due to decentralization access to information. A Productivity Method suits for assessing the value of such socio-economic enhancement of doing business with a firm</td><td>BCT's information sharing capabilities can lessen the burden of regulatory and enforcement expenses, otherwise borne by communities. Such overhead costs to the community can be estimated in monetary terms using the Damage Cost Method</td><td>BCT can provide scope for negotiation and social incentivization to achieve de-manufacturing through recycling activities. Such sustainability moves create long-term strategic advantage for organizations and enhance their macro-level standing in the region. The value accrued in monetary terms through this activity of the organization can be assessed through Calibrated and Conjoint Analysis</td><td>BCT can enable organizations to provide economic value by achieving social and environmental sustainability and help economic growth followed by increased spending and job creation. This improvement in individual welfare can be assessed by applying the Productivity Method</td></tr>
<tr><td>Socio-environmental performance</td><td>BCT operation provides a capability of tracking a system of potential social and environmental conditions that might pose environmental, health, and safety concerns in the society. The value of such data systems can be evaluated using Expenditure Approach</td><td>BCT operations can support on-going legislative requirement to comply with sustainable growth, preserving the natural setting of the ecosystem and attract travelers and tourism that benefits the community. The measurement of such organization's compliance with legislation can be captured through the Travel Cost Method</td><td colspan="2">BCT can facilitate macro-actors to collectively and securely bargain on conditions that can control damage to the environment and nature. A Damage Assessment Method can capture the value of socio-environmental damages controlled as a resultant for such bargaining by macro-actors. Additionally, a Contingent Valuation Method can be adopted to estimate the monetary incentives gained through such BCT facilitation</td></tr>
</table>

Schaltegger & Burritt, 2014; Symons & Lamberton, 2014). Factors including image and reputation, contribution to society, building community relationships, social stability, and other non-quantitative measures are difficult for organizations to integrate into their business case development portfolio (Soosay, Fearne, & Dent, 2012). The use of market or non-market valuation methods as a means of measuring performance has not been a common measurement practice for many organizations, though these techniques can provide a useful tool to measure the benefits of sustainable practices.

Blockchain technology can support various socio-environmental sustainability dimensions by addressing the "internal population," "external population," stakeholder," and "social macro" aspects. The MVM and NMVM can help organizations in the supply chain to measure the benefits of BCT, but BCT can also help gather data or evaluate the various valuation approaches. This is an interesting multiplicity and points to the various perspectives organizations can consider. The organizations who practice BCT can integrate the use of non-market valuation techniques into their performance measurements and measuring the benefits of their operation. Table 2.3 shows the integration of non-market good valuation techniques and blockchain technology and examples.

In this chapter, we have populated a classification framework as shown in Table 2.3 by integrating the social sustainability dimensions of blockchain technology and market and non-market goods valuation techniques. By using the environmental goods valuations in this format, it is possible to present the qualitative results and identify opportunities for evaluation along various social sustainability measures—highlighting the complementary and synergetic relationships between the valuation methods and blockchain technology in a socio-environmental supply chain context.

As an example, consider the internal human resources dimension of the employment stability social sustainability measures. Using Blockchain technology, the value employees would have on BCT system's decentralized information sharing mechanism can be enhanced by building trust within and between organizations. Trust is increased in a BCT supported supply chain through decentralized consensus and information sharing. The value of a trusting environment can be linked to a contingent valuation method to measure this socio-environmental sustainability measure. Many additional examples exist in the table, some are examples of how the BCT, and sustainable supply chain environment can be used to benefit the socio-environmental dimension and how this benefit can be measured.

Other examples show how BCT can support the valuation approach by gathering some data.

BCT can address the indicators of social sustainability such as *internal human resources*, as seen in Table 2.3. An organization can address on-going health and safety practices within their Blockchain technology operations and demonstrate how cost to consumers can be lowered by improving worker health. An organization can use the *averting expenditure method* to quantify the value individuals place on improving safety aspects through BCT operations. The *cost of illness method* enables organizations to quantify the value of the safety and health outcomes through a BCT; for example illness tracking—throughout a supply chain may help in epidemiological studies as when a pandemic or virus occurs, or in a food supply chain when there is contaminated goods through traceability. An organization can measure values employees place on their jobs and desirable workplaces based on BCT by evaluation and providing feedback incentives through incentivization mechanisms.

The BCT can address the human capital, productive capital, and community capital preservation aspects of social sustainability when it focuses on the "external population." Due to traceability, transparency, and traceability aspects of BCT, the organization is able to reduce the flow of waste or contamination by identifying the sources of contamination and help to preserve the human, community, and productive capital. An organization can address on-going regulatory and public services within its BCT since its activities help to identify the sources of contamination or spillages or leakages in the supply chain operation process and provide improved safety while preventing further contamination. The organization can use non-market valuation methods to estimate the benefits of BCT in this situation. The *cost-of-illness method* estimates values for morbidity based on the concept that an individual would be willing to pay at least as much as the cost of treating an illness as the cost of contracting the illness. The *cost of illness method* can be utilized to estimate the individual's willingness to pay for the organization's BCT which avoids costs to the community from altered ecosystems. The *averting expenditure method* can estimate social benefits by the change in spending on goods that are substitutes for a cleaner environment (Abrahams, Hubbell, & Jordan, 2000; Maciosek, Xin, Butani, & Pechacek, 2015).

Table 2.3 shows that the organizations can assess how stakeholder empowerment activities lead to greater awareness and participation among constituents supporting blockchain technology operations, especially for ecosystem preservation and health hazard reductions, using various valuation methods such as the contingent valuation methods, discrete choice method or expenditure approach, among others.

Table 2.3 also portrays that for macro-social issues, the environmental goods valuation methods can enable organizations to determine the importance of BCT that focuses on on-going economic welfare improvement activities. When an organization's BCT activities reduce damage to the environment or contribute to social measures then that can reduce societal damage such as reducing poverty or the loss of jobs. By using the *damage cost method*, organizations can estimate the costs the community would incur if organizations did not comply with regulatory agencies. The assessment would rate the importance area consumers place on compliance. Keeping records through BCT for all these initiatives and encouraging reporting and data gathering through incentives and building activities through negotiation systems can enhance supply chain data acquisition through BCT. This situation is especially pertinent for areas dependent on gathering environmental data that are not readily available. For example, regions of the world that have vulnerable populations that can gather data through blockchain incentive systems can greatly enhance valuation approaches. Gathering data for choice models by asking community and supply chain members is one possible approach where BCT can collect data.

Overall, many non-market goods valuation approaches can be applied through BCT from a social dimension perspective. Whether they are internal or external dimensions, organizational or societal benefits, the tools can be applied. Our examples in this section, as well as in Table 2.3, provide a more comprehensive view of resources available for companies to value and justify a business case for supply chain and Blockchain technology operations sustainability aspects.

2.7 Discussion and implications for future research

The business examples within the combined BCT and social sustainability indicators space are important to consider. Future research must gather data to support the benefits posited and provide illustrative examples for others to follow as they implement this emerging technology beyond current use.

Although BCT may have many beneficial characteristics, the applications—although feasible—may be better suited for other approaches to help gather data. Justifying the BCT in these environments may even be simpler than the necessary evaluation techniques. But, bringing all the dimensions together (1) BCT, (2) Socio-environmental dimensions, and (3) Supply chains to evaluate programs and their activities can be quite complex and not feasible with standard quantitative business evaluation techniques such as simple cost. The advantages and disadvantages can be broad and intangible, and indirect valuation methods including market and non-market valuation are needed.

Careful planning, management, and implementation are needed in these circumstances. In fact, maybe the biggest barrier to understanding and building a stronger socio-environmentally sustainable supply chain is being able to gather data and determine value to society, organizations, their supply chains, and community. We posit data gathering and efficiency improvements through BCT, and valuing through non-market and market valuation techniques.

The research questions on feasibility and limitations of these approaches need investigation. Currently, no practical example exists, although a variety of possibilities may be presented. This situation, although making it difficult to currently implement, has substantial room for further research and development. There are a number of performance measurement tools available for sustainable supply chains; extending performance evaluation through digitization through BCT holds promise. We presented examples of this promise in this chapter.

References

Abrahams, N. A., Hubbell, B. J., & Jordan, J. L. (2000). Joint production and averting expenditure measures of willingness to pay: Do water expenditures really measure avoidance costs? *American Journal of Agricultural Economics*, *82*(2), 427–437.

Brandenburg, M., Govindan, K., Sarkis, J., & Seuring, S. (2014). Quantitative models for sustainable supply chain management: Developments and directions. *European Journal of Operational Research*, *233*(2), 299–312.

Burritt, R. L., & Schaltegger, S. (2010). Sustainability accounting and reporting: Fad or trend? *Accounting, Auditing & Accountability Journal*, *23*(7), 829–846. https://doi.org/10.1108/09513571011080144.

Chen, Y. (2018). Blockchain tokens and the potential democratization of entrepreneurship and innovation. *Business Horizons*, *61*(4), 567–575.

de Sousa Jabbour, A. B. L., Jabbour, C. J. C., Foropon, C., & Godinho Filho, M. (2018). When titans meet–Can industry 4.0 revolutionise the environmentally-sustainable manufacturing wave? The role of critical success factors. *Technological Forecasting and Social Change*, *132*, 18–25.

Francisco, K., & Swanson, D. (2018). The supply chain has no clothes: Technology adoption of blockchain for supply chain transparency. *Logistics*, *2*(1), 2.

Groening, C., Sarkis, J., & Zhu, Q. (2018). Green marketing consumer-level theory review: A compendium of applied theories and further research directions. *Journal of Cleaner Production*, *172*, 1848–1866.

Hervani, A. A., Sarkis, J., & Helms, M. (2017). Environmental goods valuation for social sustainability: A conceptual framework. *Technological Forecasting & Social Change Journal*, *125*, 137–153.

Kelley, J. (2018). *Global diamond & jewelry market tracks authenticity with IBM Blockchain*. IBM. Available online https://www.ibm.com/blogs/think/2018/04/global-jewelry-ibm-blockchain/.
Kouhizadeh, M., & Sarkis, J. (2018). Blockchain practices, potentials, and perspectives in greening supply chains. *Sustainability*, *10*(10), 3652.
Kouhizadeh, M., Sarkis, J., & Zhu, Q. (2019). At the Nexus of Blockchain technology, the circular economy, and product deletion. *Applied Sciences*, *9*(8), 1712.
Kshetri, N. (2017). Can blockchain strengthen the internet of things? *IT Professional*, *19*(4), 68–72.
Maciosek, M.V., Xin, X., Butani, A. L., & Pechacek, T. F. (2015). Smoking-attributable medical expenditures by age, sex, and smoking status estimated using a relative risk approach. *Preventive Medicine*, *77*, 162–167.
Macrinici, D., Cartofeanu, C., & Gao, S. (2018). Smart contract applications within blockchain technology: A systematic mapping study. *Telematics and Informatics*, *35*(8), 2337–2354.
Mani, V., Jabbour, C. J. C., & Mani, K.T. (2020). Supply chain social sustainability in small and medium manufacturing enterprises and firms' performance: Empirical evidence from an emerging Asian economy. *International Journal of Production Economics*, *227*, 107656.
Marnewick, C., & Labuschagne, L. (2005). A conceptual model for enterprise resource planning (ERP). *Information Management & Computer Security*, *13*(2), 144–155.
Marshall, A. (1930). *Principles of economics* (8th ed.). London: Macmillan and Co. Ltd.
Mitchell, R. C., & Carson, R.T. (1989). *Using surveys to value public goods: The contingent valuation method*. Washington, D.C.: Resources for the Future, Johns Hopkins University Press.
Nakamoto, S. (2009). *Bitcoin: A peer-to-peer electronic cash system*. Unpublished manuscript, available at https://bitcoin.org/bitcoin.pdf/. accessed 10 December 2019.
Ølnes, S., Ubacht, J., & Janssen, M. (2017). Blockchain in government: Benefits and implications of distributed ledger technology for information sharing. *Government Information Quarterly*, *34*(3), 355–364.
Saberi, S., Kouhizadeh, M., & Sarkis, J. (2018). Blockchain technology: A panacea or pariah for resources conservation and recycling? *Resources, Conservation and Recycling*, *130*, 80–81.
Saberi, S., Kouhizadeh, M., Sarkis, J., & Shen, L. (2019). Blockchain technology and its relationships to sustainable supply chain management. *International Journal of Production Research*, *57*(7), 2117–2135.
Sarkis, J., & Hervani, A. A. (2017). *Sustainable supply chain performance measurement: concerns and directions*. Nova Science Publishers, Inc. Fourth Quarter, Fall.
Sarkis, J., Hervani, A. A., & Helms, M. (2013). Applying economic non market valuation for sustainable supply chain performance measurement and evaluation. In A. Lindgreen, S. Sen, F. Maon, & J.Vanhamme (Eds.), *Sustainable value chain management: Analyzing, designing, implementing, and monitoring for social and environmental responsibility* Gower Publishing. Spring.
Sarkis, J., & Zhu, Q. (2018). Environmental sustainability and production: Taking the road less travelled. *International Journal of Production Research*, *56*(1–2), 743–759. https://doi.org/10.1080/00207543.2017.1365182.
Schaltegger, S., & Burritt, R. (2014). Measuring and managing sustainability performance of supply chains: Review and sustainability supply chain management framework. *Supply Chain Management*, *19*(3), 232–241. https://doi.org/10.1108/SCM-02-2014-0061.
Schönborn, G., Berlin, C., Pinzone, M., Hanisch, C., Georgoulias, K., & Lanz, M. (2019). Why social sustainability counts: The impact of corporate social sustainability culture on financial success. *Sustainable Production and Consumption*, *17*, 1–10.
Soosay, C., Fearne, A., & Dent, B. (2012). Sustainable value chain analysis—A case study of Oxford Landing from "vine to dine". *Supply Chain Management*, *17*(1), 68–77. https://doi.org/10.1108/13598541211212212.

Svensson, G., & Wagner, B. (2015). Implementing and managing economic, social and environmental efforts of business sustainability: Propositions for measurement and structural models. *Management of Environmental Quality*, *26*(2), 195–213. https://doi.org/10.1108/MEQ-09-2013-0099.

Swan, M. (2015). *Blockchain: Blueprint for a new economy*. O'Reilly Media, Inc.

Symons, C., & Lamberton, G. (2014). Building a social case for business sustainability. *Journal of Economic and Social Policy*, *16*(2), 9.

Varsei, M., Soosay, C., Fahimnia, B., & Sarkis, J. (2014). Framing sustainability performance of supply chains with multidimensional indicators. *Supply Chain Management*, *19*(3), 242–257. https://doi.org/10.1108/SCM-12-2013-0436.

Wang, Y., Han, J. H., & Beynon-Davies, P. (2019). Understanding blockchain technology for future supply chains: A systematic literature review and research agenda. *Supply Chain Management*, *24*(1), 62–84. https://doi.org/10.1108/SCM-03-2018-0148.

Yermack, D. (2017). Corporate governance and blockchains. *Review of Finance*, *21*(1), 7–31.

CHAPTER 3

Project selection for sustainable additive manufacturing: A case study

Rohit Agrawal[a] and S. Vinodh[a]
[a]Department of Production Engineering, National Institute of Technology, Tiruchirappalli, Tamil Nadu, India

3.1 Introduction

Modern manufacturing scenario has been witnessing increased adoption of additive manufacturing (AM) technologies (Tracey, Vonderembse, & Lim, 1999). Sustainable AM has been gaining importance from the viewpoint of sustainable production systems (Aurich, Linke, Hauschild, Carrella, & Kirsch, 2013). Sustainability benefits of AM include high resource efficiency, design freedom, reduced energy consumption, generation of light-weight products, and so on (Despeisse & Ford, 2015; Rejeski, Zhao, & Huang, 2018). In this viewpoint, project selection for sustainable AM involves multiple criteria. The present study involves project selection for sustainable AM with three project alternatives and 18 criteria. Fuzzy analytic hierarchy process (AHP) is used for analyzing the alternatives. The prioritized project would enable the practitioners to deploy sustainable AM concepts effectively in the organization.

3.2 Literature review

Freitas, Almeida, Bártolo, and Bártolo (2016) analyzed environmental impact of extrusion-based additively manufactured product. They compared parts with different orientation, different density, and different layer thickness. From the comparison, they found that for different orientations, material and energy consumption were different.

Tang, Mak, and Zhao (2016) proposed an architecture for sustainable design optimization of AM processes. They compared binder jetting technology of AM process with computer numerical control (CNC) milling

Green Production Engineering and Management
https://doi.org/10.1016/B978-0-12-821238-7.00012-9

by using life cycle Assessment (LCA) approach. For comparison, they fabricated same part using both techniques and analyzed their environmental impact using LCA approach. They found that binder jetting process was more sustainable compared with CNC process as it consumes less energy and CO_2 emissions are also lesser with binder jetting process.

Despeisse, Yang, Evans, Ford, and Minshall (2017) proposed a framework for sustainable AM by roadmapping and value analysis approach. The developed framework helped in systematically analyzing the sustainability of additively manufactured product considering all stages of its life cycle.

Frăţilă and Rotaru (2017) did survey on issues associated with sustainable AM. From the survey, they concluded that energy consumption was the more dominant factor for environmental impact associated with both AM and traditional manufacturing processes. They also found that energy consumption varied with different 3D printers and also it varied with material used and pattern of fabrication.

Lundbäck and Lindgren (2017) discussed the sustainable benefits of AM. They also did finite element simulation to show the applicability of AM in minimizing wastage. Through modeling and simulation, they showed the sustainable benefits of light-weight structures fabricated using AM process.

Peng et al. (2017) aimed to compare three different manufacturing technologies, namely, conventional manufacturing, AM, and remanufacturing. They fabricated impeller using these three techniques and analyzed them. They used LCA approach for comparison. From the analysis, they concluded that remanufacturing was the most ecofriendlier alternative next to AM and conventional manufacturing.

Verma and Rai (2017) presented two major issues associated with AM processes, namely, material wastage and energy consumption. They optimized the process (layer and part levels) facilitating AM processes toward sustainability. They presented several case studies and compared AM with conventional manufacturing process to substantiate the utility of the contributed method.

Peng, Kellens, Tang, Chen, and Chen (2018) described an approach for AM to make it more sustainable. They showed that material consumption was the most prominent factor affecting the environmental impact associated with AM process. They did LCA on AM product. They identified the critical factors such as design and manufacturing parameters, end-of-life (EoL) strategies, and material preparation stage for reducing the resource consumption.

Priarone, Ingarao, Lunetto, Di Lorenzo, and Settineri (2018) aimed to compare the environmental impact of AM process with conventional manufacturing process. They have chosen jet engine part as a case part. They fabricated the case part using both AM and conventional manufacturing processes. They then used LCA approach to compare the environmental impact of part fabricated with both technologies. They concluded that AM would be more sustainable because of less material wastage and ability to produce light-weight structure.

Al-Ghamdi (2019) aimed to analyze the sustainability of AM process and to ensure minimum energy and material consumption. They analyzed FDM process parameters for optimizing specific mass and specific energy. ABS material has been considered to build parts. They used Taguchi's L46 orthogonal array for analysis purpose. They presented the optimum process parameters for printing parts for ensuring sustainable production through FDM machine.

Jiang, Liu, Li, Cong, and Zhang (2019) aimed to evaluate the sustainability of laser-based AM processes. They used energy-based LCA method to evaluate the sustainability of laser engineered net shaping (LENS) AM process. They also compared LENS with CNC machining for manufacturing gear. They computed the sustainability index of different processes and concluded that LENS process possesses more sustainability index compared to CNC machining.

Agrawal and Vinodh (2019) identified twenty factors influencing sustainable AM through literature survey. They developed structural model using total interpretive structural modeling (TISM) approach to interpret the link between factors and to identify the most dominant factors. From the analysis, they found that flexibility in manufacturing was the most dominant factor which drives all other factors. They recommended to include design-related factors in SAM for enhancing the practical applicability of the study.

3.3 Methodology

AHP is the most preferred MCDM approach for evaluating and prioritizing the alternatives. The aim of AHP is to identify and select best alternative by considering certain criteria. AHP involves a pairwise comparison of criteria to find weight of each criterion. The pairwise comparison of criteria can be done with the help of expert input. Due to uncertainty in the human thinking style, fuzzy-based AHP is recommended for the prioritization of alternatives (Huang, Chu, & Chiang, 2008). Fuzzy AHP methodology includes the uncertainty and vagueness associated with the judgment. Steps involved in fuzzy AHP are as follows:

Table 3.1 Fuzzy variables used in data collection.

Fuzzy variable	Meaning	Triangular number	Fuzzy number
EI	Equally important	(1, 1, 2)	1
MMI	Moderately more important	(2, 3, 4)	3
SMI	Strongly more important	(4, 5, 6)	5
VSI	Very strongly more important	(6, 7, 8)	7
EMI	Extremely more important	(8, 9, 10)	9

Step 1: First step is to specify the fuzzy scale used for collecting the data. The scale for data collection is being taken from the literature source (Ayağ & Özdemir, 2006). The scale is presented in Table 3.1.

The data need to be collected in the form of fuzzy numbers (1, 3, 5, 7, 9).

Step 2: Next step is to derive a pairwise comparison matrix of criteria to find the weight of each criterion.

For n criteria, $n \times n$ pairwise comparison matrix is presented below:

$$A = \begin{bmatrix} C_{11} & C_{12} & \cdots & C_{1n} \\ C_{21} & C_{22} & \cdots & C_{2n} \\ \vdots & \vdots & \ddots & \vdots \\ C_{n1} & C_{n2} & \cdots & C_{nn} \end{bmatrix}$$

Step 3: After forming the pairwise comparison matrix, the next step is to convert fuzzy number into crisp value. Here, centroid method is applied for defuzzification.

Equation used for transforming fuzzy number into crisp value is shown below:

For a fuzzy number (a,b,c), its crisp value is as follows:

$$X = \left(a + 4 * b + c\right) / 6 \tag{3.1}$$

Step 4: After developing the pairwise comparison matrix with crisp value, the next step is to develop a normalization matrix. Normalization matrix is developed to make all values lie in between 0 and 1. The normalization matrix is shown below:

$$A' = \begin{bmatrix} C'_{11} & C'_{12} & \cdots & C'_{1n} \\ C'_{21} & C'_{22} & \cdots & C'_{2n} \\ \vdots & \vdots & \ddots & \vdots \\ C'_{n1} & C'_{n2} & \cdots & C'_{nn} \end{bmatrix}$$

where

$$C'_{ij} = \frac{C_{ij}}{\sum_{i=1}^{n} C_{ij}} \tag{3.2}$$

i and j represent the row and column, respectively, and n indicate the criteria (Vinodh, Shivraman, & Viswesh, 2011).

Step 5: Next step is to calculate weights of criteria, which can be calculated using Eq. (3.2).

$$W = \begin{bmatrix} W_1 \\ W_2 \\ \vdots \\ W_n \end{bmatrix} \tag{3.3}$$

where $W_i = \dfrac{\sum_{j=1}^{n} C'_{ij}}{n}$.

Step 6: After calculating the weights of each criterion, the next step is to compare options with reference to each criteria. Pairwise comparison matrix is being made. Steps 2–5 are repeated to calculate the weight of options with reference to each criterion.

Step 7: The final step is to prioritize the alternatives by multiplying criteria weights with alternative weights for all criteria.

3.4 Case study

In the presented case study, project selection for sustainable AM has been done. Three potential projects have been considered for analysis, namely, LCA, DFAM, and DFE Guidelines application, and Materials and energy efficiency assessment. The considered three projects have the potential to enhance the sustainability of AM. Fuzzy AHP technique is used to evaluate and select the best project for SAM. For analyzing and prioritizing alternatives, 18 criteria are being considered. The considered criteria are as follows: potential for design optimization (C1), enhanced operational efficiency (C2), design freedom (C3), simplified assemblies and products (C4), design upgradation (C5), capability to generate light-weight parts (C6), improved functionality and durability (C7), safety (C8), employee acceptance (C9), regulation compliance (C10), skills enhancement

(C11), revenues (C12), implementation cost (C13), energy usage (C14), disassembly action (C15), disposal action (C16), improved resource efficiency (C17), and minimal carbon footprint (C18) (Fanai & Burn, 1997; Kroeger & Simonovic, 1997; Lindemann, Reiher, Jahnke, & Koch, 2015; Singh, Olugu, Musa, Mahat, & Wong, 2016; Wong, Lee, & Sun, 2012; Zaras, Marin, & Boudreau-Trude, 2012). A pairwise comparison matrix has been formed by collecting expert opinion for analyzing the criteria. Table 3.2 shows the pairwise comparison matrix of criteria. Eq. (3.1) is used for defuzzification. Table 3.3 depicts the pairwise comparison matrix in crisp number.

Eq. (3.2) is used to form a normalization matrix, so that all the values that lie in the pairwise comparison matrix fall in between 0 and 1. Table 3.4. shows the normalization matrix.

Weights of criteria are calculated using Eq. (3.3). Table 3.5. depicts the calculated weights of each criterion.

The next step after evaluating the criteria weights is to benchmark project options in line with the considered criteria. About 18 pairwise comparison matrices have been formed to benchmark project options in line with the 18 criteria. The pairwise comparison matrix of projects along with their weights with respect to all criteria is shown in Table 3.6A–3.6R.

After comparing alternative project for each criterion, the next step is to form a global matrix containing weights of each project with respect to each criterion. Table 3.7 shows the global matrix.

3.5 Results

The goal of the present study is to analyze the projects of sustainable AM and to prioritize the projects. Three projects are being considered for analysis, namely, LCA, DFAM & DFE Guidelines application, and Materials and energy efficiency assessment. About 18 criteria are being considered for analyzing the considered projects. Fuzzy AHP methodology is being used for analysis. The prioritization of projects is done by evaluating the global weights of each project. The global weight of each project is computed by multiplying the criteria weight with the project weight for same criteria. The prioritization and ranking of alternative projects is depicted in Table 3.8.

Table 3.2 Input data in fuzzy variables.

	Potential for design optimization	Enhanced operational efficiency	Design freedom	Simplified assemblies and products	Design upgradation	Capability to generate light-weight parts	Improved functionality and durability	Safety	Employee acceptance	Regulation compliance	Skills enhancement	Revenues	Implementation cost	Energy usage	Disassembly action	Disposal action	Improved resource efficiency	Minimal carbon footprint
Potential for design optimization	1	1/3	1	1	1	1/3	1	3	1	1	1	1	1	1	1	1	1/3	1
Enhanced operational efficiency	3	1	3	1	3	1/3	1	3	1	1/3	1	1	3	1/3	1/3	1/3	1	1/3
Design freedom	1	1/3	1	1	3	1/3	1/3	1	3	1	1	3	1/3	1	1	1	1/3	1
Simplified assemblies and products	1	1	1	1	1	1	3	3	1	3	1	3	1	1/3	1	1	1/3	1
Design upgradation	1	1/3	1/3	1	1	1	1	3	1	1/3	1	3	3	1	1	1	1/3	1
Capability to generate light-weight parts	3	3	3	1	1	1	3	3	1	1	3	3	1	1	1	3	1	1
Improved functionality and durability	1	1	3	1/3	1	1/3	1	3	1	1	3	3	1	1	1	1	1/3	1

Continued

Table 3.2 Input data in fuzzy variables—cont'd

	Potential for design optimization	Enhanced operational efficiency	Design freedom	Simplified assemblies and products	Design upgradation	Capability to generate light-weight parts	Improved functionality and durability	Safety	Employee acceptance	Regulation compliance	Skills enhancement	Revenues	Implementation cost	Energy usage	Disassembly action	Disposal action	Improved resource efficiency	Minimal carbon footprint
Safety	1/3	1/3	1	1/3	1/3	1/3	1/3	1	1	1/3	1	1	1/3	1/3	1/3	1/3	1/3	1/3
Employee acceptance	1	1	1/3	1	1	1	1	1	1	1/3	1/3	1	1/3	1/3	1/3	1/3	1/3	1/3
Regulation compliance	1	3	1	1/3	3	1	1	3	3	1	3	3	1	1/3	1/3	1/3	1	1
Skills enhancement	1	1	1	1	1	1/3	1/3	1	3	1/3	1	3	1	1/3	1/3	1/3	1	1/3
Revenues	1	1	1/3	1/3	1/3	1/3	1/3	1	1	1/3	1/3	1	1/3	1/3	1/3	1/3	1/3	1/3
Implementation cost	1	1/3	3	1	1/3	1	1	3	3	1	1	3	1	1/3	1/3	1/3	1/3	1
Energy usage	1	3	1	3	1	1	1	3	3	3	3	3	3	1	1	1	1	1/3
Disassembly action	1	3	1	1	1	1	1	3	3	3	3	3	3	1	1	1	1	1/3
Disposal action	1	3	1	1	1	1/3	1	3	3	3	3	3	3	1	1	1	1	1/3
Improved resource efficiency	3	1	3	3	3	1	3	3	3	1	1	3	3	1	1	1	1	1
Minimal carbon footprint	1	3	1	1	1	1	1	3	3	1	3	3	1	3	3	3	1	1

Table 3.3 Pairwise comparison matrix in crisp value.

	C1	C2	C3	C4	C5	C6	C7	C8	C9	C10	C11	C12	C13	C14	C15	C16	C17	C18
C1	1.00	0.33	0.86	0.86	0.86	0.33	0.86	3.00	0.86	0.86	0.86	0.86	0.86	0.86	0.86	0.86	0.33	0.86
C2	3.00	1.00	3.00	0.86	3.00	0.33	0.86	3.00	0.86	0.33	0.86	0.86	3.00	0.33	0.33	0.33	0.86	0.33
C3	1.17	0.33	1.00	0.86	3.00	0.33	0.33	0.86	3.00	0.86	0.86	3.00	0.33	0.86	0.86	0.86	0.33	0.86
C4	1.17	1.17	1.17	1.00	0.86	0.86	3.00	3.00	0.86	3.00	0.86	3.00	0.86	0.33	0.86	0.86	0.33	0.86
C5	1.17	0.33	0.33	1.17	1.00	0.86	0.86	3.00	0.86	0.33	0.86	3.00	3.00	0.86	0.86	0.86	0.33	0.86
C6	3.00	3.00	3.00	1.17	1.17	1.00	3.00	3.00	0.86	0.86	3.00	3.00	0.86	0.86	0.86	3.00	0.86	0.86
C7	1.17	1.17	3.00	0.33	1.17	0.33	1.00	3.00	0.86	0.86	3.00	3.00	0.86	0.86	0.86	0.86	0.33	0.86
C8	0.33	0.33	1.17	0.33	0.33	0.33	0.33	1.00	0.86	0.33	0.86	0.86	0.33	0.33	0.33	0.33	0.33	0.33
C9	1.17	1.17	0.33	1.17	1.17	1.17	1.17	1.17	1.00	0.33	0.33	0.86	0.33	0.33	0.33	0.33	0.33	0.33
C10	1.17	3.00	1.17	0.33	3.00	1.17	1.17	3.00	3.00	1.00	3.00	3.00	0.86	0.33	0.33	0.33	0.86	0.86
C11	1.17	1.17	1.17	1.17	1.17	0.33	0.33	1.17	3.00	0.33	1.00	3.00	0.86	0.33	0.33	0.33	0.86	0.33
C12	1.17	1.17	0.33	0.33	0.33	0.33	0.33	1.17	1.17	0.33	0.33	1.00	0.33	0.33	0.33	0.33	0.33	0.33
C13	1.17	0.33	3.00	1.17	0.33	1.17	1.17	3.00	3.00	1.17	1.17	3.00	1.00	0.33	0.33	0.33	0.33	0.86
C14	1.17	3.00	1.17	3.00	1.17	1.17	1.17	3.00	3.00	3.00	3.00	3.00	3.00	1.00	0.86	0.86	0.86	0.33
C15	1.17	3.00	1.17	1.17	1.17	1.17	1.17	3.00	3.00	3.00	3.00	3.00	3.00	1.17	1.00	0.86	0.86	0.33
C16	1.17	3.00	1.17	1.17	1.17	0.33	1.17	3.00	3.00	3.00	3.00	3.00	3.00	1.17	1.17	1.00	0.86	0.33
C17	3.00	1.17	3.00	3.00	3.00	1.17	3.00	3.00	3.00	1.17	1.17	3.00	3.00	1.17	1.17	1.17	1.00	0.86
C18	1.17	3.00	1.17	1.17	1.17	1.17	1.17	3.00	3.00	1.17	3.00	3.00	1.17	3.00	3.00	3.00	1.17	1.00

Table 3.4 Normalization matrix.

	C1	C2	C3	C4	C5	C6	C7	C8	C9	C10	C11	C12	C13	C14	C15	C16	C17	C18
C1	0.039	0.012	0.032	0.042	0.034	0.025	0.039	0.068	0.024	0.039	0.028	0.020	0.032	0.059	0.058	0.052	0.020	0.075
C2	0.118	0.036	0.110	0.042	0.120	0.025	0.039	0.068	0.024	0.015	0.028	0.020	0.113	0.023	0.023	0.020	0.052	0.029
C3	0.046	0.012	0.037	0.042	0.120	0.025	0.015	0.019	0.085	0.039	0.028	0.069	0.013	0.059	0.058	0.052	0.020	0.075
C4	0.046	0.042	0.043	0.049	0.034	0.063	0.136	0.068	0.024	0.137	0.028	0.069	0.032	0.023	0.058	0.052	0.020	0.075
C5	0.046	0.012	0.012	0.058	0.040	0.063	0.039	0.068	0.024	0.015	0.028	0.069	0.113	0.059	0.058	0.052	0.020	0.075
C6	0.118	0.108	0.110	0.058	0.047	0.074	0.136	0.068	0.024	0.039	0.100	0.069	0.032	0.059	0.058	0.182	0.052	0.075
C7	0.046	0.042	0.110	0.016	0.047	0.025	0.045	0.068	0.024	0.039	0.100	0.069	0.032	0.059	0.058	0.052	0.020	0.075
C8	0.013	0.012	0.043	0.016	0.013	0.025	0.015	0.023	0.024	0.015	0.028	0.020	0.013	0.023	0.023	0.020	0.020	0.029
C9	0.046	0.042	0.012	0.058	0.047	0.086	0.053	0.026	0.028	0.015	0.011	0.020	0.013	0.023	0.023	0.020	0.020	0.029
C10	0.046	0.108	0.043	0.016	0.120	0.086	0.053	0.068	0.085	0.046	0.100	0.069	0.032	0.023	0.023	0.020	0.052	0.075
C11	0.046	0.042	0.043	0.058	0.047	0.025	0.015	0.026	0.085	0.015	0.033	0.069	0.032	0.023	0.023	0.020	0.052	0.029
C12	0.046	0.042	0.012	0.016	0.013	0.025	0.015	0.026	0.033	0.015	0.011	0.023	0.013	0.023	0.023	0.020	0.020	0.029
C13	0.046	0.012	0.110	0.058	0.013	0.086	0.053	0.068	0.085	0.053	0.039	0.069	0.038	0.023	0.023	0.020	0.020	0.075
C14	0.046	0.108	0.043	0.148	0.047	0.086	0.053	0.068	0.085	0.137	0.100	0.069	0.113	0.069	0.058	0.052	0.052	0.029
C15	0.046	0.108	0.043	0.058	0.047	0.086	0.053	0.068	0.085	0.137	0.100	0.069	0.113	0.081	0.068	0.052	0.052	0.029
C16	0.046	0.108	0.043	0.058	0.047	0.025	0.053	0.068	0.085	0.137	0.100	0.069	0.113	0.081	0.080	0.061	0.052	0.029
C17	0.118	0.042	0.110	0.148	0.120	0.086	0.136	0.068	0.085	0.053	0.039	0.069	0.113	0.081	0.080	0.071	0.061	0.075
C18	0.046	0.108	0.043	0.058	0.047	0.086	0.053	0.068	0.085	0.053	0.100	0.069	0.044	0.208	0.205	0.182	0.071	0.088

Table 3.5 Computed criterion weights.

Criteria	Weight
Potential for design optimization (C1)	0.044
Enhanced operational efficiency (C2)	0.039
Design freedom (C3)	0.046
Simplified assemblies and products (C4)	0.062
Design upgradation (C5)	0.053
Capability to generate light-weight parts (C6)	0.077
Improved functionality and durability (C7)	0.055
Safety (C8)	0.022
Employee acceptance (C9)	0.024
Regulation compliance (C10)	0.055
Skills enhancement (C11)	0.036
Revenues (C12)	0.022
Implementation cost (C13)	0.048
Energy usage (C14)	0.076
Disassembly action (C15)	0.078
Disposal action (C16)	0.079
Improved resource efficiency (C17)	0.08
Minimal carbon footprint (C18)	0.105

Consistency check:
$\lambda_{max} = 20.732$.
$CI = (\lambda_{max} - N)/(N-1)$.
$CI = (20.732–18)/(18–1)$.
$CI = 2.732/17 = 0.139$.
$RI = 1.536$ (for $n = 12$).
$CR = CI/RI$.
$CR = 0.139/1.536$.
$CR = 0.090$.

Table 3.6A Weights of project based on criterion "Potential for design optimization".

Potential for design optimization	LCA	DFAM & DFE Guidelines application	Materials and energy efficiency assessment	Weight
LCA	1	1/3	1	0.19
DFAM & DFE Guidelines application	3	1	3	0.599
Materials and energy efficiency assessment	1	1/3	1	0.211

Consistency check.
$\lambda_{max} = 3.003$.
$CI = 0.001$.
$RI = 0.524$ (for $n = 3$).
$CR = 0.003$.

Table 3.6B Weights of project based on criterion "Enhanced operational efficiency".

Enhanced operational efficiency	LCA	DFAM & DFE Guidelines application	Materials and energy efficiency assessment	Weight
LCA	1	3	1	0.407
DFAM & DFE Guidelines application	1/3	1	1/3	0.143
Materials and energy efficiency assessment	1	3	1	0.451

$\lambda_{max} = 3.003$, CR = 0.003.

Table 3.6C Weights of project based on criterion "Design Freedom".

Design Freedom	LCA	DFAM & DFE Guidelines application	Materials and energy efficiency assessment	Weight
LCA	1	1/3	1	0.19
DFAM & DFE Guidelines application	3	1	3	0.599
Materials and energy efficiency assessment	1	1/3	1	0.211

$\lambda_{max} = 3.003$, CR = 0.003.

Table 3.6D Weights of project based on criterion "Simplified assemblies and products".

Simplified assemblies and products	LCA	DFAM & DFE Guidelines application	Materials and energy efficiency assessment	Weight
LCA	1	1/3	1	0.19
DFAM & DFE Guidelines application	3	1	3	0.599
Materials and energy efficiency assessment	1	1/3	1	0.211

$\lambda_{max} = 3.003$, CR = 0.003.

Table 3.6E Weights of project based on criterion "Design Upgradation".

Design Upgradation	LCA	DFAM & DFE Guidelines application	Materials and energy efficiency assessment	Weight
LCA	1	1/3	1	0.19
DFAM & DFE Guidelines application	3	1	3	0.599
Materials and energy efficiency assessment	1	1/3	1	0.211

$\lambda_{max} = 3.003$, $CR = 0.003$.

Table 3.6F Weights of project based on criterion "Capability to generate light-weight parts".

Capability to generate light-weight parts	LCA	DFAM & DFE Guidelines application	Materials and energy efficiency assessment	Weight
LCA	1	1/3	1	0.19
DFAM & DFE Guidelines application	3	1	3	0.599
Materials and energy efficiency assessment	1	1/3	1	0.211

$\lambda_{max} = 3.003$, $CR = 0.003$.

Table 3.6G Weights of project based on criterion "Improved functionality and durability".

Improved functionality and durability	LCA	DFAM & DFE Guidelines application	Materials and energy efficiency assessment	Weight
LCA	1	1/3	1	0.19
DFAM & DFE Guidelines application	3	1	3	0.599
Materials and energy efficiency assessment	1	1/3	1	0.211

$\lambda_{max} = 3.003$, $CR = 0.003$.

Table 3.6H Weights of project based on criterion "Safety".

Safety	LCA	DFAM & DFE Guidelines application	Materials and energy efficiency assessment	Weight
LCA	1	3	1	0.407
DFAM & DFE Guidelines application	1/3	1	1/3	0.143
Materials and energy efficiency assessment	1	3	1	0.451

$\lambda_{max} = 3.003$, CR = 0.003.

Table 3.6I Weights of project based on criterion "Employee acceptance".

Employee acceptance	LCA	DFAM & DFE Guidelines application	Materials and energy efficiency assessment	Weight
LCA	1	1	1	0.3
DFAM & DFE Guidelines application	1	1	1	0.332
Materials and energy efficiency assessment	1	1	1	0.368

$\lambda_{max} = 3.003$, CR = 0.003.

Table 3.6J Weights of project based on criterion "Regulation compliance".

Regulation compliance	LCA	DFAM & DFE Guidelines application	Materials and energy efficiency assessment	Weight
LCA	1	3	1	0.407
DFAM & DFE Guidelines application	1/3	1	1/3	0.143
Materials and energy efficiency assessment	1	3	1	0.451

$\lambda_{max} = 3.003$, CR = 0.003.

Table 3.6K Weights of project based on criterion "Skill Enhancement".

Skill Enhancement	LCA	DFAM & DFE Guidelines application	Materials and energy efficiency assessment	Weight
LCA	1	1/3	1	0.19
DFAM & DFE Guidelines application	3	1	3	0.599
Materials and energy efficiency assessment	1	1/3	1	0.211

$\lambda_{max} = 3.003$, CR = 0.003.

Table 3.6L Weights of project based on criterion "Revenues".

Revenues	LCA	DFAM & DFE Guidelines application	Materials and energy efficiency assessment	Weight
LCA	1	3	1	0.432
DFAM & DFE Guidelines application	1/3	1	1	0.214
Materials and energy efficiency assessment	1	1	1	0.354

$\lambda_{max} = 3.136$, CR = 0.13.

Table 3.6M Weights of project based on criterion "Implementation cost".

Implementation cost	LCA	DFAM & DFE Guidelines application	Materials and energy efficiency assessment	Weight
LCA	1	3	1	0.432
DFAM & DFE Guidelines application	1/3	1	1	0.214
Materials and energy efficiency assessment	1	1	1	0.354

$\lambda_{max} = 3.136$, CR = 0.13.

Table 3.6N Weights of project based on criterion "Energy usage".

Energy usage	LCA	DFAM & DFE Guidelines application	Materials and energy efficiency assessment	Weight
LCA	1	3	1	0.407
DFAM & DFE Guidelines application	1/3	1	1/3	0.143
Materials and energy efficiency assessment	1	3	1	0.451

$\lambda_{max} = 3.003$, CR = 0.003.

Table 3.6O Weights of project based on criterion "Disassembly action".

Disassembly action	LCA	DFAM & DFE Guidelines application	Materials and energy efficiency assessment	Weight
LCA	1	1/3	1	0.19
DFAM & DFE Guidelines application	3	1	3	0.599
Materials and energy efficiency assessment	1	1/3	1	0.211

$\lambda_{max} = 3.003$, CR = 0.003.

Table 3.6P Weights of project based on criterion "Disposal action".

Disposal action	LCA	DFAM & DFE Guidelines application	Materials and energy efficiency assessment	Weight
LCA	1	1	1	0.3
DFAM & DFE Guidelines application	1	1	1	0.332
Materials and energy efficiency assessment	1	1	1	0.368

$\lambda_{max} = 3.003$, CR = 0.003.

Table 3.6Q Weights of project based on criterion "Improved resource efficiency".

Improved resource efficiency	LCA	DFAM & DFE Guidelines application	Materials and energy efficiency assessment	Weight
LCA	1	3	1	0.407
DFAM & DFE Guidelines application	1/3	1	1/3	0.143
Materials and energy efficiency assessment	1	3	1	0.451

$\lambda_{max} = 3.003$, CR = 0.003.

Table 3.6R Weights of project based on criterion "Minimal carbon footprint".

Minimal carbon footprint	LCA	DFAM & DFE Guidelines application	Materials and energy efficiency assessment	Weight
LCA	1	3	1	0.407
DFAM & DFE Guidelines application	1/3	1	1/3	0.143
Materials and energy efficiency assessment	1	3	1	0.451

$\lambda_{max} = 3.003$, CR = 0.003.

3.6 Conclusions

Sustainable AM is gaining much importance from the viewpoint of developing ecofriendly designs coupled with resource and material efficiency, optimal geometries, and simplified products. In this viewpoint, the selection of appropriate projects is of vital importance. Three projects, namely, LCA (P_1), DFAM & DFE Guidelines application (P_2), and Materials and energy efficiency assessment (P_3), are being considered with the 18 criteria. Fuzzy AHP is applied to solve the MCDM problem. The priority order is DFAM & DFE Guidelines application, Materials and energy efficiency assessment, and LCA. The prioritized project would enable the practicing engineers to deploy projects for enhanced performance from the viewpoint of sustainability and AM.

Table 3.7 Global matrix.

Project	C1	C2	C3	C4	C5	C6	C7	C8	C9	C10	C11	C12	C13	C14	C15	C16	C17	C18
P1	0.19	0.407	0.19	0.19	0.19	0.19	0.19	0.407	0.3	0.407	0.19	0.432	0.432	0.407	0.19	0.3	0.407	0.407
P2	0.599	0.143	0.599	0.599	0.599	0.599	0.599	0.143	0.332	0.143	0.599	0.214	0.214	0.143	0.599	0.332	0.143	0.143
P3	0.211	0.451	0.211	0.211	0.211	0.211	0.211	0.451	0.368	0.451	0.211	0.354	0.354	0.451	0.211	0.368	0.451	0.451

Table 3.8 Ranking of alternative projects.

Project	Global weight	Rank
LCA	0.3	3
DFAM & DFE Guidelines application	0.373	1
Materials and energy efficiency assessment	0.327	2

References

Agrawal, R., & Vinodh, S. (2019). Application of total interpretive structural modelling (TISM) for analysis of factors influencing sustainable additive manufacturing: A case study. *Rapid Prototyping Journal*, *25*(7), 1198–1223.

Al-Ghamdi, K.A. (2019). Sustainable FDM additive manufacturing of ABS components with emphasis on energy minimized and time efficient lightweight construction. *International Journal of Lightweight Materials and Manufacture*, *2*(4), 338–345.

Aurich, J. C., Linke, B., Hauschild, M., Carrella, M., & Kirsch, B. (2013). Sustainability of abrasive processes. *CIRP Annals*, *62*(2), 653–672.

Ayağ, Z., & Özdemir, R. G. (2006). A fuzzy AHP approach to evaluating machine tool alternatives. *Journal of Intelligent Manufacturing*, *17*(2), 179–190.

Despeisse, M., & Ford, S. (2015, September). The role of additive manufacturing in improving resource efficiency and sustainability. In *IFIP international conference on advances in production management systems* (pp. 129–136). Cham: Springer. https://doi.org/10.1007/978-3-319-22759-7_15.

Despeisse, M., Yang, M., Evans, S., Ford, S., & Minshall, T. (2017). Sustainable value roadmapping framework for additive manufacturing. *Procedia CIRP*, *61*, 594–599. https://doi.org/10.1016/j.procir.2016.11.186.

Fanai, N., & Burn, D. H. (1997). Reversibility as a sustainability criterion for project selection. *The International Journal of Sustainable Development & World Ecology*, *4*(4), 259–273.

Frăţilă, D., & Rotaru, H. (2017). Additive manufacturing—a sustainable manufacturing route. In *94. MATEC web of conferences* (p. 03004). EDP Sciences. https://doi.org/10.1051/matecconf/20179403004).

Freitas, D., Almeida, H.A., Bártolo, H., & Bártolo, P.J. (2016). Sustainability in extrusion-based additive manufacturing technologies. *Progress in Additive Manufacturing*, *1*(1–2), 65–78.

Huang, C. C., Chu, P. Y., & Chiang, Y. H. (2008). A fuzzy AHP application in government-sponsored R&D project selection. *Omega*, *36*(6), 1038–1052.

Jiang, Q., Liu, Z., Li, T., Cong, W., & Zhang, H. C. (2019). Emergy-based life-cycle assessment (Em-LCA) for sustainability assessment: A case study of laser additive manufacturing versus CNC machining. *The International Journal of Advanced Manufacturing Technology*, *102*(9–12), 4109–4120.

Kroeger, H. I., & Simonovic, S. P. (1997). Development of a risk measure as a sustainable project selection criterion. *The International Journal of Sustainable Development & World Ecology*, *4*(4), 274–285.

Lindemann, C., Reiher, T., Jahnke, U., & Koch, R. (2015). Towards a sustainable and economic selection of part candidates for additive manufacturing. *Rapid Prototyping Journal*, *21*(2), 216–227.

Lundbäck, A., & Lindgren, L. E. (2017). Finite element simulation to support sustainable production by additive manufacturing. *Procedia Manufacturing*, 7, 127–130. https://doi.org/10.1016/j.promfg.2016.12.033.

Peng, T., Kellens, K., Tang, R., Chen, C., & Chen, G. (2018). Sustainability of additive manufacturing: An overview on its energy demand and environmental impact. *Additive Manufacturing*, *21*, 694–704. https://doi.org/10.1016/j.addma.2018.04.022.

Peng, S., Li, T., Wang, X., Dong, M., Liu, Z., Shi, J., et al. (2017). Toward a sustainable impeller production: Environmental impact comparison of different impeller manufacturing methods. *Journal of Industrial Ecology*, *21*(S1), S216–S229.

Priarone, P. C., Ingarao, G., Lunetto, V., Di Lorenzo, R., & Settineri, L. (2018). The role of re-design for additive manufacturing on the process environmental performance. *Procedia CIRP*, *69*, 124–129. https://doi.org/10.1016/j.procir.2017.11.047.

Rejeski, D., Zhao, F., & Huang, Y. (2018). Research needs and recommendations on environmental implications of additive manufacturing. *Additive Manufacturing*, *19*, 21–28. https://doi.org/10.1016/j.addma.2017.10.019.

Singh, S., Olugu, E. U., Musa, S. N., Mahat, A. B., & Wong, K.Y. (2016). Strategy selection for sustainable manufacturing with integrated AHP-VIKOR method under interval-valued fuzzy environment. *The International Journal of Advanced Manufacturing Technology*, *84*(1–4), 547–563.

Tang, Y., Mak, K., & Zhao, Y. F. (2016). A framework to reduce product environmental impact through design optimization for additive manufacturing. *Journal of Cleaner Production*, *137*, 1560–1572. https://doi.org/10.1016/j.jclepro.2016.06.037.

Tracey, M., Vonderembse, M. A., & Lim, J. S. (1999). Manufacturing technology and strategy formulation: Keys to enhancing competitiveness and improving performance. *Journal of Operations Management*, *17*(4), 411–428.

Verma, A., & Rai, R. (2017). Sustainability-induced dual-level optimization of additive manufacturing process. *The International Journal of Advanced Manufacturing Technology*, *88*(5–8), 1945–1959.

Vinodh, S., Shivraman, K. R., & Viswesh, S. (2011). AHP-based lean concept selection in a manufacturing organization. *Journal of Manufacturing Technology Management*, *23*(1), 124–136.

Wong, T. N., Lee, L. H., & Sun, Z. (2012). CSR and environmental criteria in supplier selection. *Computers and Operations Research*, *34*(12), 3516–3540.

Zaras, K., Marin, J. C., & Boudreau-Trude, B. (2012). Dominance-based rough set approach in selection of portfolio of sustainable development projects. *American Journal of Operations Research*, *2*(04), 502.

CHAPTER 4

Does priority sector lending affect profitability? An empirical study on Indian public sector banks

Sudin Bag[a], Nilanjan Ray[b], and Biswajit Roy[c]
[a]Department of Business Administration, Vidyasagar University, Midnapor, West Bengal, India
[b]Marketing and Examination Coordinator, Adamas University, Kolkata, West Bengal, India
[c]Future Business School, Kolkata, West Bengal, India

4.1 Introduction

Priority sectors are those of the economy that seldom get timely and requisite attention in the absence of special dispensation to them. It is considered as highly vulnerable sector. It indicates special assistance is emergence for the development in all aspects. After 1969, priority sector lending and social banking concepts came into mainstream and were adopted for the credit development (Gupta & Kumar, 2004). That's why, the adaptation of priority sectors concept was adopted with the aim of catering the bank lending. That reflects the lending of credit to each bank with national priority.

For being the apex bank of India, RBI has given due importance to providing credits to alleviate these vulnerable sectors. RBI has recognized agriculture, micro, small, and medium enterprises (MSME), education, export credit, renewable energy, and housing as some of the remarkable priority sectors in India. Though initially, the priority sector was among only few sectors, including agriculture, small business, and other small industries that were neglected sectors since long, and for the development through bank credit, they were treated as priority sector (Bag & Islam, 2017). But the concept of priority sector has gone through several changes, and since then, several new sectors and areas have been incorporated within the purview of this sector. The list of priority sector lending in India has been updated over the period of time as per the annual policy statement of RBI for the year 2005–06. It is also stated that the eligibility criteria have been enlarged to provide space for several new sectors as the priority sector.

Green Production Engineering and Management
https://doi.org/10.1016/B978-0-12-821238-7.00008-7

The scope and extent of priority sector lending (PSL) has been amended several times, and a remarkable change to include several sectors under the umbrella of priority sector is done. Though it is apprehended that it will dilute the definition of the priority sector with the focus on the needy sectors of the economy, weaker section of society will get completely lost. In these circumstances, to examine the need for emergence of priority sector lending prescriptions, an internal working committee has been set up in the Reserve Bank of India which has been chaired by Shri C.S. Murthy. The committee also reviewed the existing policies on priority sector lending, including the wide range of segments constituting the priority sector lending targets and subtargets and recommending changes for the betterment of performance, if any, required in this regard. Based on the draft, technical paper submitted by the internal working group, and the feedback received thereon, the guidelines were revised.

On the basis of the recommendations of the internal working committee of RBI, the guideline for priority sector lending (PSL) was revised on April 30, 2007, and overall target was fixed at 40% of net bank credit (NBC) for domestic banks (including PSBs and private banks) and 32% for foreign banks. The target for the agricultural advances was fixed at 18% for domestic banks, and the target in the case of small-scale industries was fixed at 12% for foreign banks (RBI, Department of Banking Supervision, 1999).

4.1.1 Categories of priority sector advances

The broad categories of Indian priority sector for all scheduled commercial banks are discussed in the following sections.

4.1.2 Agriculture and allied activities

Agriculture advances include short-, medium-, and long-term loans given for agriculture and allied activities directly to individual farmer, self-help groups (SHGs), or joint liability groups (JLGs) of individual farmers without limit and to others (such as corporate partnership firms and institutions) up to Rs. 20 lakhs for taking up agriculture and allied activities.

4.1.3 Small-scale industries (direct and indirect finance)

Advances to small-scale industries (SSIs) include loans given to SSI units, which are engaged in manufacture, processing, or preservation of goods and whose investment is in plant and machinery (original cost) excluding land and building.

4.1.4 Small business/service enterprises

It includes small business, retail trade, professional and self-employed persons, small road and water transport operators, and other enterprises.

4.1.5 Microcredit

Provision of credit and other financial services and products of very small amounts not exceeding Rs. 50,000 per borrower to the poor in rural, semiurban, and urban areas, either directly or through a group mechanism, for enabling them to improve their living standards, constitutes microcredit.

4.1.6 Education loans

Education loans include loans and advances granted to individuals for educational purpose, up to Rs. 10 lakhs for studies in India and Rs. 20 lakhs for studies abroad and do not include those granted scholarship by the institutions.

4.1.7 Housing loans

Loans up to Rs. 15 lakhs for construction of houses by individuals (excluding loans granted by banks to their own employees) and loans for repairs to the damaged houses of individuals up to Rs. 1 lakh in rural and semiurban areas and up to Rs. 2 lakhs in urban areas are given.

It is important to note that for the purposes of priority sector lending, adjusted net bank credit (ANBC) or credit equivalent of off-balance sheet exposures (CEOBE) imply NBC plus investments made by banks in non-SLR bonds held in HTM category (Table 4.1).

Table 4.1 Different categories of Priority sector in Indian Public Sector Banks.

Categories of priority sector	Targets/subtargets of Indian PSBs
Total agricultural advances	18% of ANBC or CEOBE, whichever is higher
SSI advances	Overall priority sector target of 40% of ANBC or CEOBE, whichever is higher
Microenterprises within SSI	60% of SSI advances
Export credit	Export credit is not a part of priority
Advances to weaker sections	10% of ANBC or CEOBE, whichever is higher
Differential rate of interest scheme	1% of total advances outstanding as at the end of the previous year
Total priority sector advances	40% of ANBC or CEOBE, whichever is higher

4.2 Literature review

RBI, Department of Banking Supervision (1999) studied the impact of priority sector advances on NPAs and found that the incidence of NPAs in priority sector is much higher in view of the fact that priority sector advances constitute 30%–32% of the gross bank credit. Vyas Committee (2001) also observed that commercial banks seem to have shied in extending rural credit as they are dealing with vast number of small accounts. Niranjana and Anbumani (2002) observed that there is anxiety among the bankers that the advances to priority sector resulted in a loss of interest income due to highly subsidized lending rates. In the context of enormous demand for funds for priority sector, Shete (2002) examined priority sector advances by PSBs during the postreform years and found that the PSBs are not able to reach the prescribed target of lending to priority sector.

Bhati (2006) examined the lending climate for banks in emerging economies like India. The study found that due to the government policies, banks in India undertake many additional risks when they lend. This study focused on one specific aspect of lending relationship between branch manager and loan officer of bank branches in India, and recommended social risk evaluation is therefore more appropriate for risk evaluation and reduction by banks in India. Rikta (2006) in a policy note on institutional lending and financing policy for small sector in Bangladesh examined the effectiveness of financial, fiscal, and related policies for financing the small sector. The study suggested that the financial institutions need to have the financial capacity to accept the lending risks along with access to appropriate funding, which fulfill the clients' requirements. An assessment on the causes and consequences of NPAs of commercial banks, by Rajesham and Rajender (2007), concluded that a strong political basis will only be able to find satisfactory solution to the problem of mounting NPAs. Sooden and Kumar (2007) analyzed the priority sector lending in the postreform period and suggested for a balanced approach as development agent and sustainability of directed lending keeping in view the quantum of NPA in priority sector advances.

4.3 Objectives of the study

This study aims to gain insights into the relationship between profitability and priority sector advances by public sector banks (PSBs). The following broad objectives are laid down for the purpose of the study:

1. To study the trends of priority sector lending by the public sector banks in India.
2. To measure the relationship between profitability of public bank and priority sector lending in Indian public sector banks

4.3.1 Hypothesis of the study

To achieve the objective of the present study, the researcher has formulated and tested the following null hypothesis:

Ho:There is no significant relationship between profitability of the public sector banks in India and priority sector lending.

Ha: Otherwise, it is true.

4.4 Research methodology

For the purpose of the study, priority sector lending of the public sector bank and profitability, a total of 21 Indian public sector banks have been taken into account. To measure the relationship between priority sector lending and profitability of the public sector banks, secondary data (Panel) have been taken into consideration. The relevant data for the study have been collected from the annual report of Reserve Bank of India. Apart from the RBI annual report, the necessary data have been sought from the Capitaline database.

To assess the relationship between the profitability and the priority sector lending, return on assets (ROA) and return on equity (ROE) have been considered as the measure of the profitability of public sector banks which are dependent variables of the study. ROA indicates the profit earned per unit of assets and most importantly shows the management capability to utilize the bank's financial as well as real investment resources to maximize profits. However, ROA of banks depend on the bank's policy decision and other uncontrollable issues regarding the economy and regulatory framework of the government. ROE, on the other hand, indicates the effectiveness and efficiency of bank management toward the utilization of shareholders fund. In the literature on banking, the profitability of bank is most commonly represented by one or both of the two alternative measures: the ratio of net profit to equity, ROE, and the ratio of net profit to assets, ROA (Altunbas & Ibanez, 2004; Athanasoglou, Delis, & Staikouras, 2006; Bag & Omrane, 2020; Berger & Humphrey, 1997; De Young & Rice, 2004; Fries & Taci, 2004; Ganesan, 2001; Gul, Irshad, & Zaman, 2011). We have considered ROA and ROE as the major indicator of banks profitability. Thus, selection of dependent variables of our study is justified.

On the other side, priority sector lending is treated as the independent variable of the study. It is difficult to interpret that only one factor influences on the profitability of the banking sector. Therefore, size of the bank (lag value of the total assets), net nonperforming assets, net sales, assets turnover ratio, equity to total asset, total cost (operating cost), and operating cost ratio are taken as the control variables in the study. The period of the study is ranging between March 2008 and March 2019, i.e., 12 years. To assess the relationship between the profitability and the priority sector lending, fixed effects and random effects models have been used in Eviews-10.

4.5 Findings

Table 4.2 represents the changes of priority sector lending sectorwise. It is very remarkable that the growth of lending is very positive and development of lending in priority sector is positively upward rising (Table 4.3). It is also clear that after the postreform period, scope and extend of various new priority sectors have emerged with the existing sectors.

The additional lending of priority sector target of 2% of adjusted net bank credit (ANBC) each year from 2015–16 to 2019–20 has to be achieved by lending to sectors other than exports.

Table 4.2 Changes of under priority sector lending sectorwise.

Items	Target/subtarget (% of ANBC/OBE)	Amount (%) (Amount in ₹ billion)	
		March 08	March 18
Total priority sector advances		1641 (42.5)	20,723 (39.9)
Agriculture		586 (17.1)	9321 (18)
Small and marginal farmers		469 (13.7)	4633 (8.9)
Noncorporate individual farmers		–	6647 (12.8)
Microenterprises		154 (12.2)	3317 (6.4)
Weaker sections		–	5946 (11.5)

Table 4.3 Growth of percentage of total priority sector lending.

Financial year	The total priority sector as percentage of ANBC or credit equivalent amount of off-balance sheet exposure, whichever is higher
2015–16	32
2016–17	34
2017–18	36
2018–19	38
2019–20	40

In order to measure the relationship between profitability of Indian public sector banks and priority sector lending, we have performed regression analysis. The data are panel in nature; therefore, it is important to select the effect to analyze the result. Initially, we have performed Hausman test to understand the effects of model. The Hausman test fundamentally perceives the predictor variables or endogenous regressors in a regression model (RM). In this study, Hausman test is done to identify the predictor variables are endogenous or not. In the case of our panel data analysis, the Hausman test is utilized to select the random or fixed effects model, where, H0 defines that the chosen model is random effects and H1 defines that the chosen model is fixed effects.

From the analysis of Hausman test using LS (LS and AR) method, the Prob. value ($p = 0.0234$) showcases that the chosen model will be fixed effects model ($p < 0.05$). In this study, the fixed effects model is applied to find out the relation between ROA and priority sector lending in Indian banks. Result of Hausman test is demonstrated in Table 4.4.

After Hausman test analysis, we further run the fixed effects model to assess the relation between ROA and priority sector lending (PSL). The result reveals that R^2 value is 0.814 indicating 81% variation in the baking profitability in terms of return of assets is explained by the independent variables (Table 4.5). It is further confirmed by the F statistics which is

Table 4.4 Result of correlated random effects—Hausman test.

Equation: Untitled
Test period random effects

Test summary	Chi-Sq. statistic	Chi-Sq. d.f.	Prob.
Period random	19.218061	9	0.0234

** WARNING: Estimated period random effects variance is zero.
Period random effects test comparisons:

Variable	Fixed	Random	Var(Diff.)	Prob.
APS	0.000003	0.000003	0.000000	0.7368
ATR	0.154469	0.091030	0.002406	0.1959
EPU	0.000018	−0.000024	0.000000	0.0032
NNPA	0.000003	0.000007	0.000000	0.5676
OC_TA	−0.516472	−0.525823	0.000741	0.7312
SIZE	0.055477	−0.037212	0.002001	0.0383
TC	0.000003	−0.000003	0.000000	0.6177
TOTAL_ASSETS	−0.000000	−0.000000	0.000000	0.7524

statistically significant. This implies that the independent-variable priority sector lending and other control variables such as assets turnover ratio, equity paid-up, net nonperforming assets, net sales, operating cost to total assets, size, and total cost are the good determinants of dependent variable, viz. return on assets (ROA). However, the result reveals that the *p* value of independent variable, viz. priority sector lending, is 0.286 which is not statistically significant. Thus, the null hypothesis of the study is accepted in this case. Therefore, it seems that there is no statistical significant relationship

Table 4.5 Result of relationship between ROA and priority sector lending

Period random effects test equation:

Dependent variable: ROA
Method: Panel least squares

Sample (adjusted): 3/01/2008 3/01/2019

Periods included: 12
Cross sections included: 20

Total panel (balanced) observations: 240

Variable	Coefficient	Std. error	t-Statistic	Prob.
C	−0.360642	1.301825	−0.277028	0.7820
APS	2.79E-06	2.61E-06	1.068160	0.2866
ATR	0.154469	0.062809	2.459351	0.0147
EPU	1.85E-05	3.90E-05	0.473041	0.6367
NNPA	3.31E-06	1.11E-05	0.299719	0.7647
OC_TA	−0.516472	0.059506	−8.679282	0.0000
SIZE	0.055477	0.092087	0.602440	0.5475
TC	2.84E-06	1.98E-05	0.143489	0.8860
TOTAL_ASSETS	−3.45E-07	6.07E-07	−0.568003	0.5706

Effects Specification

Period fixed (dummy variables)

R-squared	0.814154	Mean dependent var	0.357375
Adjusted R-squared	0.797182	S.D. dependent var	0.908920
S.E. of regression	0.409335	Akaike info criterion	1.134867
Sum squared resid	36.69458	Schwarz criterion	1.439423
Log likelihood	−115.1841	Hannan-Quinn criter.	1.257581
F-statistic	47.96978	Durbin-Watson stat	1.421804
Prob(F-statistic)	0.000000		

between ROA as the indicator of profitability of Indian banks and the lending to the priority sector.

We have performed the similar statistical analysis to assess the relationship between ROE another important measure of bank profitability and priority sector lending in Indian banks (Table 4.6). From the analysis of Hausman test using LS (LS and AR) method, the Prob. value ($p = 0.0234$) showcases that the chosen model will be fixed effects model ($p < 0.05$). In this study, the fixed effect model is applied to find out the relation between ROE and other variables.

After Hausman test analysis, we further run fixed effects model to assess the relation between ROE and priority sector lending. Similar to Table 4.5, Table 4.7 demonstrates the significant result between the dependent and independent variable. The R^2 value is 0.769, which reflects that the dependent variable is well explained by the independent variable (Bag, Aich, & Islam, 2020) and the result of F-statistics indicates that the independent variable along with other control variables is the sound determinant of the dependent variable. On the other side, the result also similarly revealed as the previous relationship that there is no significant relationship between ROE and priority sector lending in Indian banks.

Table 4.6 Correlated random effects—Hausman test on ROE.

Equation: Untitled
Test period random effects

Test summary	Chi-Sq. statistic	Chi-Sq. d.f.	Prob.
Period random	64.200269	11	0.0000

Period random effects test comparisons:

Variable	**Fixed**	**Random**	**Var(Diff.)**	**Prob.**
APS	−0.000006	0.000016	0.000000	0.2445
ATR	−0.560615	−3.576177	0.412440	0.0000
EPU	0.002301	0.002449	0.000000	0.6329
ETA	−2.890331	−3.714151	0.259465	0.1058
NET_SALES	0.000233	0.000304	0.000000	0.3438
NNPA	−0.000250	−0.000670	0.000000	0.0000
OC_TA	−1.270333	−1.826260	0.200207	0.2141
SIZE	−1.728730	−4.262715	0.635365	0.0015
TC	−0.000058	0.000560	0.000000	0.0000
TOTAL_ ASSETS	−0.000007	−0.000026	0.000000	0.0002

Table 4.7 Result of relationship between ROE and PSL.

Period random effects test equation:

Dependent variable: ROE
Method: Panel least squares

Sample (adjusted): 3/01/2008 3/01/2019

Periods included: 12
Cross sections included: 20

Total panel (balanced) observations: 240

Variable	Coefficient	Std. error	t-Statistic	Prob.
C	37.16644	22.16308	1.676952	0.0950
APS	−6.14E-06	4.02E-05	−0.152553	0.8789
ATR	−0.560615	0.958282	−0.585021	0.5591
EPU	0.002301	0.000964	2.386334	0.0179
ETA	−2.890331	1.759493	−1.642707	0.1019
NET_SALES	0.000233	0.000217	1.073874	0.2841
NNPA	−0.000250	0.000145	−1.727865	0.0854
OC_TA	−1.270333	0.955612	−1.329339	0.1851
SIZE	−1.728730	1.609149	−1.074314	0.2839
TC	−5.78E-05	0.000273	−0.212200	0.8322
TOTAL_ASSETS	−6.86E-06	1.65E-05	−0.415536	0.6782

Effects specification

Period fixed (dummy variables)

R-squared	0.765902	Mean dependent var	10.41371
Adjusted R-squared	0.742168	S.D. dependent var	10.54226
S.E. of regression	5.353060	Akaike info criterion	6.284139
Sum squared resid	6218.190	Schwarz criterion	6.617700
Log likelihood	−731.0967	Hannan-Quinn criter.	6.418540
F-statistic	32.27093	Durbin-Watson stat	1.510189
Prob(F-statistic)	0.000000		

4.6 Conclusion

It is observed during the study that the growth of lending is positively inclined. As a result, it is also found that in postreform period, various new priority sectors have emerged. A Hausman test helped to identify the study as fixed affect model that is applied to find out the relation between ROA and priority sector lending in Indian banks. The associated regression model analysis revealed that the priority sector lending and other control variables such as assets turnover ratio, equity paid-up, net nonperforming assets, net sales, operating cost to total assets, size, and total cost are the good determinants of return on assets (ROA). But no significant relationship was found between ROA as the indicator of profitability of Indian banks and the lending to the priority sector. Similarly, when the Hausman test is applied to establish the relationship between ROE another important measure of bank profitability and priority sector lending in Indian banks, the study again indicated toward the application of fixed effects model. Though the associated regression model only could identify a significant relationship between ROE and priority sector lending in Indian banks in spite of the good indication (with R^2 value of 0.769) that independent variable along with other control variables is the sound determinant of dependent variable. This study can provide an indication to the future researchers about the select of model for different indicators of bank profitability and priority sector lending in Indian banks.

At the nutshell, we can conclude that the general myth regarding the relationship between the profitability of Indian public sector banks and priority sector lending is not statistically correct. As our general knowledge, we have a conception that the priority sector lending is the main hindrance of profitability of Indian public sector banks but results of our study demonstrated that profitability of the Indian PSBs is not significantly affected by the priority sector lending. Thus, there might be other determinants influenced by the bank's profitability. So, banks may consider the priority sector lending as one of the important sectors for economical as well as social development in India.

References

Altunbas, Y., & Ibanez, D. M. (2004). Merger and Acquisitions and Bank Performance in Europe: The Role of Strategic Similarities. *European Central Bank Working Paper, 398*, 1–37.

Athanasoglou, P. P., Delis, M. D., & Staikouras, C. K. (2006). Determinants of Bank Profitability in the South Eastern European Region. *Bank of Greece Working Paper, 47*, 1–37.

Bag, S., Aich, P., & Islam, A. (2020). Behavioral intention of "digital natives" toward adapting the online education system in higher education. *Journal of Applied Research in Higher Education*. https://doi.org/10.1108/JARHE-08-2020-0278.

Bag, S., & Islam, S. (2017). Non-performing assets a biggest challenge in banking sector—A comparative study between India and Bangladesh banking sector. *ICTACT Journal on Management Studies, 3*(4), 620–625. https://doi.org/10.21917/ijms.2017.0084.

Bag, S., & Omrane, A. (2020). Corporate social responsibility and its overall effects on dinancial performance: Empirical evidence from Indian companies. *Journal of African Business*. https://doi.org/10.1080/15228916.2020.1826884.

Berger, A. N., & Humphrey, D. B. (1997). Efficiency of Financial Institutions: International Survey and Directions for Future Research. *European Journal of Operational Research, 98*, 175–212.
Bhati, S. (2006). Trust between Branch Managers and Loan Officers of Indian Banks. *International Review of Business Research Papers, 2*(4), 51–58.
De Young, R., & Rice, T. (2004). NonInterest Income and Financial Performance at U.S. Commercial Banks. *The Financial Review, 39*, 101–127.
Fries, S., & Taci, A. (2004). Cost efficiency of Banks in Transition: Evidence from 289 Banks in 15 Post-communist Countries. *EBRD Working paper, 86*, 1–29.
Ganesan, P. (2001). Determinants of Profits and Profitability of Public Sector Banks in India: A Profit Function Approach. *Journal of Financial Management and Analysis, 14*(1), 27–37.
Gul, S., Irshad, F., & Zaman, K. (2011). Factors affecting bank profitability in Pakistan. *The Romanian Economic Journal, XIV*(39), 61–87.
Gupta, S., & Kumar, S. (2004). Dimensions and Prospects of Non-performing Assets: Challenges Before the Banking Sector Reforms in the New Millennium. In *Edited Book Banking in the New Millennium* (pp. 279–291).
Niranjana, S., & Anbumani, V. (2002). *Social Objectives and Priority Sector Lending, Banking and Financial Sector Reforms in India* (pp. 231–233). Deep and Deep Publications.
Rajesham, C., & Rajender, K. (2007). Management of NPAs in Indian Scheduled Commercial Banks. In *The Chartered Accountant of India.*
RBI, Department of Banking Supervision. (1999). Some aspects and issues relating to NPAs in commercial banks. *RBI Bulletin, LIII*(7), 913–930.
Rikta, N. N. (2006). Institutional Lending and Financing Policy for SMEs in Bangladesh. In *Policy Note, Federation of Bangladesh Chamber of Commerce and Industry, Dhaka.*
Shete, N. B. (2002). *Priority Sector Advances By Public Sector Banks during the Post Reform Period (1992-93 to 2000-01).* Working Paper, NIBM.
Sooden, M., & Kumar, S. (2007). Priority sector lending in post reform period. *Finance India, XXI*(4), 1389–1404.
Vyas Committee. (2001). *Report of the Expert Committee on Rural Credit (V S Vyas Committee).* Mumbai: NABARD.

CHAPTER 5

The *momentum* for green human resources' management

Diana Fernandes[a] **and Carolina Machado**[b]
[a]University of Minho, Braga, Portugal
[b]School of Economics and Management, University of Minho, Braga, Portugal

5.1 Introduction

Roome (1992: 11) states that the "critical turning point for society, business, and environment" has emerged in the 1990s, once environmental management issues had stood up as a major concern, in need of deep strategic thinking and action planning for the years ahead. According to Vlados, Deniozos, and Chatzinikolaou (2018: 21), the "idiosyncratic track of the current crisis of globalization" is imprinted, directly, "in the deepest level of the socioeconomic evolution of every region and every agent of action on the planet," so that it "penetrates and dissects the real life of every individual, group, and the entire organization, their plans and actions of every kind." So, in the contemporary context, organizations cannot ignore environmental issues; indeed, and as mentioned by Walton, Handfield, and Melnyk (1998: 2), we currently assist to an "[i]ncreasing government regulation and stronger public mandates for environmental accountability."

Following this, the milieu to the choices the society faces over the environment has been set out in the literature on the connection between conservation and development, uttering that we are currently moving toward environmentally sustainable approaches, which require an ultimate reassessment of our activities and organizations. Based on this, already back in the 1990s, Roome (1992: 11) stated that "two critical aspects" could be clearly seen in the process and the problem of "adjusting to uncertain environmental futures—these involve recognition and transformation." The first one "describes the need for society and individuals to acknowledge that their current lifestyles and activities contribute to the environmental change on a significant, potentially life-threatening scale." Therefore, advocated that a "radical shift in how society views and understands the world's resources"

https://doi.org/10.1016/B978-0-12-821238-7.00003-8

was imperative to standstill the trend, the "acceptance of an environmental ethic to underpin individual and social behavior" bases this approach. The second one comprises the responses that follow recognition, "as pathways to more sustainable forms of resource use are sought out, identified, and implemented."

Environmental decision-making is extremely complex due to the systems' interdependent and complex dynamics and to the competing interests of multiple stakeholders. Hence, Khalili e Duecker (2013: 188) stated that proactive environmental management initiatives "are inherently multiobjective processes that require joint considerations of environmental, industrial, economic, and social criteria in all stages of decision-making." As a result, an approach to actively think about and act toward them requires a transparent decision-making process and is flexible to changing circumstances and inclusive of heterogeneity.

Literature on environmental management often advocates that, considering organizations as the main cause of environmental problems, they shall (consequently) assume a large part in addressing such issues and act toward them. Accordingly, we increasingly assist to a more frequent callout and to a wide range of eco-initiatives launched by organizations to address environmental management concerns as one organizational response to environmental degradation, supported on a developing green human resource management model.

So, with this essay, we seek to weave a theoretical reflection around the influence of social, economic, financial, political, environmental, and other external forces on human resource management approaches. We intend to launch a discussion on how they are currently marked by the vector of change and how they are increasingly oriented toward sustainability in order to contribute to the improvement of the social and environmental performance of organizations, adding the value of social relevance, calling out for a green human resource management frame. Therefore, we intend to focus on human resource management policies and practices that support the paradigm shift in organizational and people management around environmental issues. The model presented in Fig. 5.1 schematizes this. In methodological terms, this work is based on reviewing literature and empirical data of the most relevant contributions in the topic; thus, it assumes a fundamentally theoretical nature.

The contributions of this article are threefold: first, to survey and draw together the elements of environmental management; second, describing the background for environmental management, to map to the emergence of green human resource management; and third, to outline some avenues for potential further research in green human resource management. In order to do so, the article is divided into three sections: the first describes the background for environmental management, presenting change as the neuralgic concept to it; the second dissertates on the *momentum* for green human resource management, characterizing this approach and presenting some of its practices and processes by reviewing the state-of-the-art on it; and the third foresees the upcoming debates on green human resource management. So, this is a descriptive work, mapping the terrain for a theoretical framework and reviewing the existent empirical research on green human resource management.

Relevance of this essay would be responding to calls in the literature to integrate environmental management and human resource management as a subject of research (Jabbour & Santos, 2008; Jackson, Renwick, Jabbour, & Muller-Camen, 2011) and to expand the scope of strategic human resource management to incorporate sustainability and environmental issues (Osland & Osland, 2007; Wilkinson, Hill, & Gollan, 2001).

From the present article, we can conclude that the following processes and practices become vital to foster the skills among the "green" firms' employees: strict recruitment and selection strategies based on environmental consciousness, an evaluation process transfiguring into an appraisal and reward systems comprising environmental performance, as well as training and empowerment programs on this topic. For today's organizations, it is crucial to promote a proper alignment of human resource management principles with green management; the greater the strength of green human resource policies, the greater the companies' intensity to adopt environmental management systems and policies, under the premise of fostering prosperity beyond economic growth, inherently referring to *sustainability* as imperative. While some answers concerning the future of human resource management seem to be known and shared, much more is still unknown, so further research is needed.

5.2 Change as the neuralgic concept

The concept *change* comprises a strategic dimension since it refers to an important and fundamental development, referring indeed to the movement of a company from its current situation to the desired future. From this, we can derive the premise that adapting to change is the key to survival; so, organizations should prepare for the creation of complexity to cope up with the environment's complexity. Therefore, managing change has become the core in seeking the final component of successfully managing strategy, process, people, and culture in most contemporary organizations.

In fact, the ability to change rapidly, efficiently, and almost continually is a major dilemma for organizations in today's rapidly changing environment (Lorenzi & Riley, 2003). Therefore, globalization and its interdependent complex worldwide dynamics are forcing agents to make dramatic adjustments and improvements, at an increasing speed. Today, we act toward a fluid global environment, marked by constant and increasing heterogeneity, liquidity, and uncertainty (Vlados et al., 2018). A variety of complex dynamics (economic, financial, social, political, institutional, demographic, cultural, religious, climatic, geopolitical, and geostrategic) are mingled and redefined to a notable level (Avant, Finnemore, & Sell, 2010). Undeniably, "globalization can be conceived to be born and reproduced by the diverse social and economic interconnections and flows that it creates and exploits" (Avant et al., 2010: 10). Consequently, the environmental challenge is growing awareness in all agents' daily routines, being *sustainability* the motto for today's world—the "green" mindset.

As a result of globalization and its imprint, the current world is paved with increasing uncertainty and heterogeneity, which requires organizations to make changes (Chiang, 2010). Change comes at two levels. First, fundamental change requires changing culture or identity, based on the premise that sticking to an identity and denying the upcoming transformational forces may lead to failure in creating added value. Second, capacity for change implies that an agent shall be capable of responding quickly to what is occurring around.

In the current frame of heightened scrutiny of corporate behavior (Basu & Palazzo, 2008; Waddock, 2000) and increasing demand for corporate social responsibility programs (Lenssen et al., 2009), organizations are being increasingly called out as both the main cause of environmental problems, as well as the main agents to cause awareness on the topic and to mobilize and implement solutions. There is a wide range of factors influencing a company's decision to adopt an environmental strategy, specifically the financial performance, stakeholder pressure, and corporate image (Cherian & Jacob, 2012).

Not disregarding that win-win situations are possible in some organizations and contexts, we assume that for the majority of organizations, it will not be so easy to create simultaneously economic and financial efficiency and ecological, social, and human sustainability without a core change in their business strategy and organizational culture (Ehnert & Harry, 2012). In fact, knowing the pace and unpredictability of change does not mean that organizations have learned to manage change; change redefines risk, this premise implying that agility becomes more important than accuracy in reducing risk. Hence, change will lead to the need of redefining the organization's performance away from merely cutting cost to profitable growth, a fact that by itself will also require a change from human resources' theoretical approaches, practices, processes, and professionals' characteristics.

Human resource management approaches are being redesigned, innovative strategies tested and implemented, advanced information systems installed, structures reshaped, and processes reengineered. All these changes, as Yeung, Woolcock, and Sullivan (1996) point, call, however, for refurbished roles and competencies among such professionals. In fact, their skills shall imperatively align with the changing human resource management visions, strategies, structures, systems, and processes (Yeung et al., 1996). Therefore, in the current context, developing a green human resource management model of people management practice is emerging as one organizational response to environmental degradation.

5.3 The *momentum* for green human resources' management

Much recent interest has been paid to environmentalism globally. As the current trend, it is generally acknowledged that protecting, improving, and managing environmental resources is crucial. Indeed, a large number of organizations, institutions, and stakeholders (frequently with competing objectives) are responsible for policy analysis, regulatory decision-making, and priority setting for environmental actions and assessments. In fact, organizations have to currently consider themselves as playing a major role in ecological footprint reduction and give importance to social and environmental factors along with economical and financial ones, so that they can be enabled to succeed in the corporate sector (Cherian & Jacob, 2012).

This resulted in the advent of a new strategic maneuver called *green management*. Indeed, in the management field, we assist to an increase of empirical research and theoretical literature on green marketing (Peattie, 1992), green accounting (Owen, 1992), green retailing (Lai, Cheng, & Tang, 2010), and green management topics (McDonagh & Prothero, 1997). As a result, there has been an intensification in the development and use of formal scientific approaches to assist in environmental management decision-making (Renwick, Redman, & Maguire, 2012). This approach has *sustainability* as its core.

We can conceptualize *green management* as the strategy an organization adapts to conceive and act toward environmental management, in order to face environmental challenges. Its ground base is as stated by Daily and Huang (2001): organizations essentially need to balance the industrial growth and ensure the environment where one lives is well preserved and promoted. Some of the advantages associated with the adoption of this approach include a reduction in overall cost (Hart, 1997), promotion of teamwork (Jabbour, Santos, & Nagano, 2010), improvements related to firms' operational performance (Jackson et al., 2011), as well as enhancements in organizational culture (Jabbour et al., 2010).

However, in comparison, green human resource management research, defined as the human resource management aspects of environmental management, is rather diverse and fragmentary. Jackson et al. (2011: 99) state that "environmental sustainability is generating increased concern among business executives, governments, consumers, and management scholars." Nevertheless, human resources management scholars and practitioners "have been relatively slow to engage in the ongoing discussions and debates."

We can conceive green human resource management as the area of study where human resource management policies and practices are used to ensure the efficient use of natural resources in any business—in this concept lies the constant encouragement to the cause of ecology and sustainability (Ashraf, Ashraf, & Anam, 2015). Also important to highlight is that the term green human resource management refers to the "promotion of sustainable employee practices" having "every employee" has a crucial "interface" (Ahuja, 2015: 50).

Hence, it involves two essential elements: human resource practices characterized by a friendly imprint on the environment and the conservation of knowledge capital (Ahuja, 2015). So, its approach includes recruitment and selection, compensation and reward schemes, performance management, training and development, as well as employee involvement and participation, so that its main benefits occur in reduced utility costs, decreased environmental impact, improved public image, tax benefits, and increased business opportunities.

Already back in 1996, Yeung et al. set the pace for discussion pointing out the human resource management competencies for the future, dissertating on the keys to sustaining transformation of human resource management functions. Then, the book by Wehrmeyer (1996), titled *Greening People: Human Resources and Environmental Management,* highlights among the very early academic contributions to this field. Later then, Renwick et al.'s (2012) paper titled *Green human resource management: a review and research agenda* presented an exhaustive review of around 60 books and articles on the topic, published between 1988 and 2008. Another remarkable contribution has been provided by Jackson et al. (2011), presenting the state-of-the-art on the green human resources topic, as well as its future directions. Another input to the high point is the work of Cherian and Jacob (2012), reviewing the study of green human resource practices and their effective implementation in the organizations. This relates then to the analysis done by Kramar (2014), shedding light on the topic *Beyond strategic human resource management: is sustainable human resource management the next approach?*

Aware of that debate, Wagner (2013) analyzed whether benefits arising for human resource management, from environmental management activities, drive environmental management system implementation. Useful to develop that line of discussion is the later works of Opatha and Arulrajah (2014), Ahmad (2015), Ashraf et al. (2015) and then Tariq, Jan, and Ahmad (2016), who developed later a work on the same root, schematizing a conceptual model as a base to a theoretical approach on the green human resource management approach. Later then, Tang, Chen, Jiang, Paille, and Jia (2018), inspired by the fact that previous studies on green human resource management were mainly positioned at a theoretical or qualitative level, developed a valid instrument to measure such an approach offering more insights into the implication of it on individual and organizational performance.

Human resource management is currently under a continuous and sharpened scrutiny, which is prone to have a positive impact in terms of theoretical approaches and practical improvements. In fact, it as we have known needs to change to face contemporary challenges. In an active response to the pace of the current uncertainty and change, we shall keep in mind that change outside human resource management will require change within it as well and that human resource management is at an intersection in its ability to deal with this change. Thus, changing human resource management will represent important challenges and will require new competencies from such professionals, as well as reviewing established practices, processes, and theoretical approaches. We can, therefore, state that the incorporation of environmental objectives and strategies into the organization's overall strategic development goals will help it achieve an effective environmental management system (Haden, Oyler, & Humphrey, 2009).

5.3.1 The path toward a green human resource management approach

Fostering a green human resource management approach in an organization shall start with the setup of several initiatives to enhance the organization's environmental knowledge, in order then to improve its environmental management potential. Hence, we present in this section highlights to implement a green human resource management approach and review the literature on each of them. The model presented in Fig. 5.1 schematizes this.

5.3.1.1 Attracting and developing talented staff

Attracting high-quality employees is a key human resource management challenge in the current (and even more intense, as per the future expectations) "war for talent." Ehnert and Harry (2012) suggest that a fraction of employers, particularly major multinational companies, are, therefore, increasingly adopting green human resource management practices as a form of "employer branding," conscious of an increasingly environmentally aware younger generation being the current and upcoming workforce base. This would aim at improving their selection attractiveness for them. Mandip (2012) states that a company needs to promote its corporate culture and environment in an active way when it selects and recruits new employees. Based on this, they also highlight that it is essential to integrate the "green environment" mindset into the recruitment process. Doing so, this process would aim at providing information about the organizations' policies and culture to new employees and monitoring competent people to join the organization to ensure long-term benefits and avoid expenses. Hence, the goal is transfiguring it into a fair process, with every candidate having equal rights so that there are no biases grounded on idiosyncrasies, such as gender, race, age, language, disability, nationality, etc.

Indeed, and according to Renwick et al. (2012), job seekers (especially the workforce above described) prefer organizations where they perceive a close linkage between their and the organization's values. Aware of this pattern, organizations have adapted to the pace of change so that currently the recruiting organization's environmental reputation is increasingly prominent in recruitment efforts. This aligns with signaling theory in recruitment and selection, given the fact that, due to incomplete information about such process, candidates use organizational attributes (mostly environmental image and reputation) to draw hints about the company's future intentions and actions. This follows what Bauer and Aiman-Smith (1996) suggested already in the late 1990s, in an empirical study considering a fictitious firm,

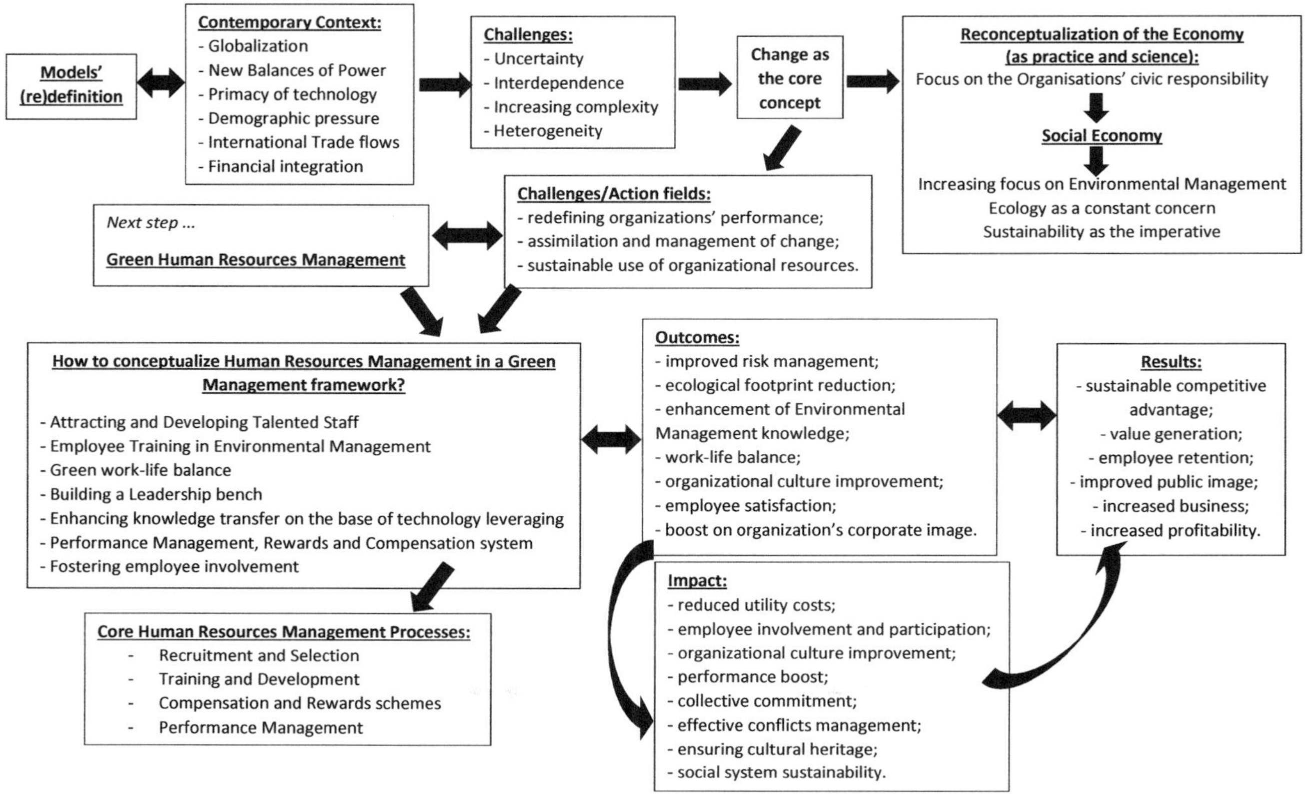

Fig. 5.1 The *momentum* for green human resources management. *Source: Adapted and expanded from Ashraf, F., Ashraf, I., & Anam, W. (2015). Green HR for businesses. International Journal of Academic Research in Business and Social Sciences, 5(8), 149–156.*

studying the effect of a pro-environmental stance on recruitment efforts, collecting evidence that when an organization presents a proactive stand toward environment protection and sustainability, its attractiveness as an employer in recruitment processes would improve.

Indeed, the *green agenda* appears to be increasingly impacting the criteria some employers require in new hires. Jabbour et al. (2010) found empirical evidence for Brazilian firms with ISO14001 certification, proving that recruiters prefer candidates with environmental knowledge and motivation. Grolleau, Mzoughi, and Pekovic (2012) corroborate this, once their work identified that the company's environmental standards impact recruitment; their data indicates that a company's environmental commitment adds to the company's profile. Focusing on the importance of contemporary environmentalism in the recruitment process in leading companies in the United States, Dechant and Altman (1994) emphasized the importance of employee perception of a firm's environmental behavior, evidence proving it is vital since employees are willing to work in an organization only when they feel it adds to their value profile. Additionally, candidate preferences for *green organizations* also seem to be impacting the organizational practice, once, as stated by Brockett (2006, p. 18) some employers are being increasingly influenced by "green job candidate" thinking in planning their recruitment strategies.

In line with the above, surveys and experimental designs by Behrend, Baker, and Thompson (2009) and Backhaus, Stone, and Heiner (2002) in the United States report university students as being attracted to work for organizations with pro-environmental images. Following this line, Brekke and Nyborg (2008) identified, modeling a theoretical approach, how companies attracted responsible employees, stating that when there is equilibrium between socially responsible and nonresponsible organizations, motivation among new recruits would be higher toward *green organizations*. Besides, if wages were similar in both companies, applicants who approach socially responsible organizations would be much higher than those who approach nonresponsible ones.

Although the literature so far presents few systematic studies on *green* recruitment practices, Renwick et al. (2012) point to a rising advice industry of self-help guides on how to find a *green* job, which includes a case study and employer interview evidence about their hiring practices. This literature reports the use of job descriptions and personnel specifications, emphasizing the job's environmental aspects and interview protocols, with this also enquiring the applicant's environmental knowledge, values, and beliefs to see whether it is corresponding (see also Cassio & Rush, 2009; Llewellyn & Golden, 2008; Parks & Helmer, 2009).

5.3.1.2 Employee training in environmental management

Training is widely seen in the literature as a key green human resource management intervention tool, having a threefold impact: to heighten staff awareness on the environmental impact of their organization's activities (Bansal & Roth, 2000); to equip staff with core environmental skills (May & Flannery, 1995); and to raise the level of "eco-literacy" and environmental expertise in the organization (Roy & Thérin, 2008).

As foreseeing and managing environmental impact is, nowadays, the major responsibility for employers and employees, organizations are increasingly acknowledging that taking into account the employees' tacit knowledge in environmental management is nuclear in identifying sources of environmental problems, managing emergency situations, and developing preventative solutions (Boiral, 2002). Indeed, to take tacit environmental knowledge into account, organizations "must challenge the predominance of formal knowledge in the management of environmental problems and promote a climate of learning that encourages the recognition and sharing of employees' experiences" (Boiral, 2002: 291).

Hence, developing an accurate and broadened environmental knowledge base is vital to the effectiveness of learning and training on this topic. As Boiral (2002: 292) points out, even it has once been perceived "as the duty of a specialized department, the control of environmental impacts is now viewed as the responsibility of all employees, who are called on to integrate these issues into their daily activities." Some organizations have, likewise, taken their employees' learning and daily commitment and turned them into the main thrust of their environmental policy.

Kunes (2001) defends that organizations that implement environmental prevention practices have as well invested in training their employees, therefore, can apply their skills toward more advanced forms of environmental management. This idea suggests, according to Darnall and Edwards Jr (2006) that an organization's environmental management capabilities "may be path-dependent in that internal proficiencies and learning accrue over a period of time," so that, aiming to attain a higher level of environmental competency, "organizations may first need to be expert in basic levels of environmental competency and acquire the socially complex or process-based resources to do so" (on the same line of argumentation, see also Christmann, 2000; Henriques & Sadorsky, 1996; Russo & Fouts, 1997).

In line with this, self-report study of managers in China (Fryxell & Lo, 2003) reveals that environmental knowledge and values predict personal environmental behaviors. Therefore, developing a learning and training framework for creating and sustaining a pro-environmental organization also requires the organization itself to hire employees who have environmental awareness and knowledge and are willing to engage in environmental management activities.

The literature so far predicts both direct and moderating effects of employee training on the organization's environmental performance. Ji, Huang, Liu, Zhu, and Cai (2012), testing among manufacturing firms in China, obtained empirical evidence that supports the relationship between a firm's environmental attitude and its performance in sustainable development, this being fed by efforts on employee training. The results suggest that employee training should moderate positively the relationship between a firm's environmental attitude and its performance in sustainable development so that with better employee training on the topic, an organization's correct environmental attitude should be more likely to help improve its performance in sustainable development.

Therefore, we call out a learning and training framework covering all employees in an organization, not only focused on the new joiners, and such development programs shall include social and environmental issues, stressing the importance of environmental management. We share the idea of Ashraf et al. (2015), alerting that important in this training, foremost when addressed to new employees, would be to link and to constantly inform on the organization's *green* policies and procedures, as well as on its (*green*) vision and mission.

Therefore, the importance of employee training in environmental management derives from the premise that well-trained and environmentally aware employees are ideally placed to identify and best manage such issues. Following this, Fernández, Junquera, and Ordiz (2003) find that a pro-environmental approach requires increased employee awareness, knowledge, and skills in both processes and materials and that this requires integrated training in environmental management to create emotional involvement in environmental concerns.

Also important to keep in mind is the fact that advanced environmental management approaches are seen to be "people-intensive" and dependent upon skill development through employee training (del Brío, Fernández, & Junquera, 2007: 494). Milliman and Clair (1996) indeed provide a very useful contribution in regard to this, alerting for several concerns which arise in regard to the use of training and development in environmental management. These comprise the need to counter employee cynicism on the relevance of the involved issues, given the fact that cynicism often arises since such training is occasionally delivered in an extremely "politically correct way," with an over-emphasis on enforcement (Rees, 1996). Thus, we suggest that organizations shall not only develop more training in environmental management but also cautiously assess the general effectiveness of it (Perron, Côté, & Duffy, 2006). Other training concerns are ensuring that managers release staff for training, as well as integrating training results into appraisal and performance management systems (Wehrmeyer, 1996).

5.3.1.3 Green work-life balance

Work-life balance can be defined as "a state where an individual tries to manage work activities with one's personal life so that there can be benefits for both organizations as well as employees" (Ashraf et al., 2015: 151). Work-life balance might be difficult to manage due to many constraints in the workplace, from which we highlight long working hours, stress derived from work overload or work environment, among others. Therefore, to control these factors, organizations shall develop policies that enable employees to manage their personal lives with their work.

Ashraf et al. (2015) call out for green human resource management as the best response to this, based on the premise of Muster and Schrader (2011), postulating that *green* work-life balance is the settlement of private and work life comprising environmental values and attitudes. The same authors advocate, indeed, that green human resource management "can meet its full potential only by considering employees in their twofold role as producers and consumers," based on the argument that "reciprocal interactions between working life and private life occur" (2011: 140). So, a "green work-life balance concept" is suggested to boost environmental behavior in those both domains. This concept's relevance lies in the fact that it provides chances in a threefold way: to the environment, to the company, and to its employees. Therefore, Ashraf et al. (2015) advocate that green human resource practices (improving organizational culture and environment, car sharing, home office, tasks done by telecommunicating, providing other benefits apart from wages—for example, wellness programs, reducing working hours, decreasing workload, delegating tasks, and implementing mentor circles and/or counseling) can be provided in organizations.

As pointed by Datta (2015) as well as per Muster and Schrader (2011), such initiatives increase work motivation and job retention, but it is also important to keep in mind that some challenges may also arise, such as employees' reluctance to allow corporate influence on their private life.

5.3.1.4 Building a leadership bench

To face the contemporary context and its challenges, it is crucial to rethink how leaders will need to lead in the future (in every management level, we may find in the organization's hierarchic chart), which skills will they be required to develop and to make use of and what subsequent human resource management implications are supporting the development of such leaders. Late in the 1990s were these concerns already foreseen (Ulrich, 1997).

To provide responses to the above questions, in a green human resource management frame, the leadership bench itself shall be *green*. To provide more insights into this, Egri and Herman (2000) conducted a study to develop a preliminary model for environmental leadership, using primary data from Canadian and American leaders, revealing that their personal values were more ecocentric, open to change and self-transcendent than other managers in different types of organizations. Following the same line, also Bansal and Roth (2000) collected empirical evidence that personal values influence *green* leadership behaviors, studying British and Japanese companies; they state that single individuals tend to be a titleholder of ecological responses, with their own values in the backend of such decision processes, rather than observing the existence of a widely applied decision rule. The geographical scope to stress this trend is wide; Branzei, Ursacki-Bryant, Vertinsky and Zhang (2004: 1075) found empirical evidence also in China to prove that leaders' cognition shapes their organizations' responses to deteriorating environmental circumstances—the results suggest that executives tend to" champion new initiatives following personal values and principles."

Building this leadership framework and connecting it to the organization's setup of a green agenda also require initiatives to actively develop empowerment and engagement in this leadership bench. Therefore, commitment from the organization's management to environmental management systems is seen as providing the underpinning framework to the organization's response to environmental management issues. Nevertheless, it is crucial to notice that, without wider employee engagement, its success may be limited. So, in today's world, leaders shall be developed to succeed as an example so that they can foster their teams to join them on the pace to fulfill the organization strategy—in this case, it is green agenda.

Rather than management only acting seeking to ensure mere employee compliance with environmental management systems, the need now is to win employees'/teams' "hearts and minds" to the environmental cause. By including and involving them in environmental management actions through green human resource management, workforce will be motivated to take ownership on environmental management issues and increasingly actively provide their contribution to a more sustainable organization. Thus, the premise that is "management initiative without employee involvement is useless," vividly pointed by Denton (1999).

There is considerable evidence that supportive managerial and supervisor behaviors in environment initiatives are significant in fostering employee engagement in environmental management. Siebenhüner and Arnold (2007) observed the learning processes of managers in medium- and large-sized German and Dutch organizations, revealing a participatory leadership style being used, in which leaders stand as active in involving employees in sustainability processes. Furthermore, Ramus and Steger (2000) dedicated to employee "eco-initiatives," conceived as any action an employee takes perceived to improve the company's environmental performance. They found a strong relationship between managerial behaviors (such as competence building, communication, rewarding, and recognizing employees) and their engagement with innovative environmental activities.

5.3.1.5 Enhancing knowledge transfer on the basis of technology leveraging

Knowledge transfer considers that best practices are shared within a firm, among firms, and between firms, and the government. Therefore, effectively managing existing knowledge stocks and transferring them between partners contribute to the organization's capabilities that differentiate its goods and services from its competitors. Thus, since it also helps organizations to provide greater value to customers, it contributes as well to the development of a competitive advantage. Hence, proactively building and managing knowledge are decisive steps to ensure the organization's survival (Collins & Hitt, 2006). As suggested by Ulrich (1997), human resource professionals must, then, master tools of learning such as knowledge generation (experimentation, benchmarking, continuous improvement, and competence acquisition) and generalization (moving information, skills, decision-making, and reward across boundaries).

In fact, Collins and Hitt (2006) state that knowledge transfers within the organization and between partners have become increasingly important with globalization, which crystallized the ferocious competition based on knowledge, on its hand turning successful knowledge transfer a significant challenge since the geographical boundaries are even more elapsed, due to the virtual era's triumph (Coff, Coff, & Eastvold, 2006). However, the knowledge management process is complex especially due to the difficulty involved in conveying tacit knowledge. Tacit knowledge has, thus, the strategic significance given its potential to be valuable, rare, and hard for rivals to imitate or acquire, sustaining a competitive advantage in the organization (Barney, 1991).

Once valuable tacit knowledge is identified, firms must transfer and replicate it to increase the scale and meet the demand for the scarce resource (Tsai, 2001), which may imply face-to-face communication and protracted learning curves (Teece, 1977). Coff et al. (2006: 465) propose that information technology might be posed as a solution in leveraging tacit knowledge. Their proposal indeed represents "a very different research agenda for the information technology literature," suggesting that "technology might be applied to leverage knowledge without codification," based on the premise that, in today's globalized world, there is a "need to leverage expertise across geographically dispersed units." In order to do so, they call out for the use of telecommunication solutions, given the fact that these can convey tacit knowledge across great distances. This argument can be framed into a green human resource management framework and fortified on the premise that "[t]he more traditional database-driven knowledge management systems appear to require more codification than this and may ultimately be of somewhat less strategic value" (Coff et al., 2006: 465). Therefore, being such applications viable, the technology's strategic role will be enhanced—so, as a supporting tool, human resource management can lie on technology, as it will change how work is done in general and how human resource management is practiced in particular.

5.3.1.6 Performance management, reward, and compensation system

Even though empirical studies have been rather limited on this topic, the literature indicates a positive effect of reward systems in the organization's environmental performance (Chinander, 2001; Daily, Bishop, & Steiner, 2007; Epstein & Roy, 1998; Fernández et al., 2003; Forman & Jorgensen, 2001; Laabs, 1992; Ramus, 2001). As a general trend, such evidence demonstrates that most organizations do not formally reward improved environmental performance (Chinander, 2001; Daily et al., 2007;Denton, 1999 ; Kitazawa & Sarkis, 2000). As pointed by Denton (1999), the ones that encourage environmental improvement efforts infrequently mainly attach monetary incentives to environmental performance.

Nevertheless, the literature has been reporting that involving employees in environmental management improves the key outcomes of environmental management systems, which embrace reducing waste (May & Flannery, 1995); efficient and sustainable resource usage (Florida & Davison, 2001); and reducing pollution from workplaces (Denton, 1999; Kitazawa & Sarkis, 2000). Therefore, we can remark that employee involvement in environmental management prints effects through three core processes: tapping employees' tacit knowledge obtained via their close links to the production process (Boiral, 2002); appealing, committing, and empowering employees to make suggestions for environmental enhancements (Govindarajulu & Daily, 2004); and by nurturing a workplace culture supportive on environmental management improvement efforts.

The literature states clearly that market, business, and regulatory demands remain as the main drivers of environmental management. Nevertheless, it also vividly notes that employees themselves are often reported as a major source of pressure in organizations so that they address and actively act toward environmental issues (Berry & Rondinelli, 1998). Both perspectives coexist, there is a symbiosis rather than an exclusive logic. This climaxes the argument that wider employee participation in environmental management (rather than blindly curbing involvement to managers and specialists) is central to achieve successful outcomes (Remmen & Lorentzen, 2000). Therefore, the organization's performance management system shall include environmental management issues and evaluate different departments/teams/employees' results on the efforts taken in that direction. This shall also deeply connect to a reward system and compensation scheme that comprises their environmental management potential.

Performance management is a procedure of improving individuals' and teams' capabilities so that the organization's sustainable competitive advantage can be achieved. It supports the premise that human capital provides an organization's competitive advantage (Reynolds & Ablett, 1998), so that the people composing the organization are its vital edge (Band, Scanlan, & Tustin, 1994). Therefore, there shall be a continuous development of performance management systems to better associate employees' competencies with the organization's goal, with the critical aim to enhance the organizations' performance. Thus, the goal of the performance management system is to motivate and emphasize values and behaviors (Risher, 2003). In green human resource management, performance management systems can be used to indicate how the performance standards can be achieved at different organization levels, as well as to provide useful data on the environmental performance of a department, a team, or a single employee.

The linkage between environmental management and reward/compensation management for other than senior management is rarely reported in the literature (Renwick et al., 2012). As a trend, Bashford (2008), reviewing the association between reward scheme and environmental management performance for this type of organization staff, reports a "dearth of activity."

Ramus (2001) presents a useful contribution, giving more insights on competency-based reward schemes for frontline staff acquiring specific environmental competencies (knowledge of environmental legislation, as an example). The author argues that, since those skills have been perceived by the organization as an important advantage once allowed to better anticipate emergency situations and proactively address them, such employees are seen to help organizations stop serious environmental accidents.

Other than that, financial incentives have been introduced into the company's environmental management reward strategies. As so, employers are progressively reviewing their reward scheme and employment condition policies and practices, critically evaluating if they align with environmental objectives. In practice, and as noted by Renwick et al. (2012), this has mainly focused on benefits such as transport rather than pay itself.

There is some developing evidence that rewarding the workforce for their environmental management performance on the basis of paying is effective. For example, Fernández et al. (2003) report companies with contingent remuneration for senior managers having higher environmental management performance than those with fixed salaries. Nevertheless, it is noteworthy to highlight that the causation for such positive impact is not clearly addressed in such studies. As Renwick et al. (2012) point, it may be that organizations are only reacting to environmental performance concerns by implementing managerial reward for environmental management performance, which drives to a rewarding pattern with a subtle "enforcement" attitude in its backend. Indeed, Russo and Harrison (2005) present empirical evidence based on the electronic industry in the United States, suggesting that managerial pay results from (rather than causes) environmental performance, which, by its end, boosts firms to remain reactive on environmental issues. Stanwick and Stanwick (2001) report that American firms on the Forbes' list reveal a strong relationship between CEO compensation (total compensation and salary) and firm environmental reputation, even though CEOs are not necessarily rewarded for their organization's environmental management record and moreover are not stimulated toward doing so by their organization's compensation systems. Berrone and Gomez-Mejia's (2009) link between environmental performance and executive compensation in American firms, strongly supporting environmental performance as positively associated with CEO total pay.

Nevertheless, not only financial reward schemes have a positive impact on this dynamic. Organizational practice on linking environmental management and reward schemes for employees below senior management indeed mostly concentrates on giving employees nonmonetary recognition reward for their environmental management performance (Govindarajulu & Daily, 2004). Ramus (2001) shed light on the importance of nonfinancial incentives on environmental policy implementation, surveying employees of European companies, observing that, in cases where there was perceived supervisory support and better acknowledgment as of praise, there has also been an increased employees' commitment in upholding environmental policies. So, reward such as recognition (like daily praise and company awards) are seen to have a significant impact on employee willingness to generate eco-initiatives. Ramus (2001: 93) calls out the importance of such compensation schemes on the premise that those initiatives feed an open style of communication, which heartens employees to discuss their environmental ideas "in an honest and unrestrained manner."

Govindarajulu and Daily (2004) empirically relate a company's environmental performance and its relationship to employee motivation, identifying that management commitment, employee empowerment, reward, recognition, feedback, and review assume as key factors to be addressed in an organization's human resource management policy so that effective green management practices can be successfully promoted and fulfilled. The importance of such management is, on the same line, stressed by Taylor (1992), whose empirical evidence states that, when companies offered an incentive as *green* awards, employees were more inclined to follow principles of green management. Also, Forman and Jorgensen (2001) report similar data; analyzing employees' participation in environmental work programs in several organizations, they state that reward or compensation has been provided to employees to undertake the additional responsibility of participating in environmental efforts.

Other than these nonfinancial reward, Simms (2007) also reports annual award dinners to recognize exemplary behaviors of employees and employers in environmental management initiatives. On the same line, Govindarajulu and Daily (2004) call out paid vacations, time off, and gift certificates as important innovative nonmonetary reward for employee environmental management actions. In fact, Renwick et al. (2012) point out that these recognition-based reward for staff (shedding light for example on company-wide public recognition), used in large American companies, are offered at different levels (for example by CEOs annually for an individual, team, and divisional contributions to waste reduction, company-wide team excellence awards, among others) and in nontraditional forms (for example, awarding employees with opportunities to attend green events).

Important to notice from the above argumentation is that the core of an organization's reward system and compensation scheme lies the quest for employee retention. Denton (1999) states that employees who are satisfied and loyal to their organizations put more effort into their work, so they can add more value to its betterment. So, if an organization adopts *green* practices, it will be able to retain employees in a more successful tax once it will be providing them the opportunity to balance personal and professional values and attitudes, thus becoming *green* individuals as well as *green* employees.

So, the above thinking calls out the need for an environmental reward strategy, even though it has been clear in the literature that a reward system is not common in environmental management. Even though, Massoud, Daily, and Bishop (2008: 17) alert that "the characteristics of environmental issues make traditional individual reward systems difficult," given the fact that environmental management is fairly recent and organizations have so far neglected to define reward for environmental management efforts; the scope of defining new environmental performance standards is prone to be so wide-ranging so that it may cost the organization an undue amount of time to do so; and the cross-boundary nature of environmental issues problematizes rewarding on an individual basis (Daily, Bishop, & Govindarajulu, 2009). Peacock (1993) as well as Young and Welford (1998) also shed light on the complexity of environmental performance measurements, which turns difficult to reward improved environmental performance in financial terms. Thus, Rothenberg (2003) states that environmental issues shall rely on a multidisciplinary, team-oriented approach to solving problems.

5.4 Foreseen debates in a green human resources' management approach

The argumentation above still leaves space for a set of questions ahead for human resource management. To desiccate them, we follow the logic of Ulrich (1997), already predicting those challenges in the late 1990s, based on the upcoming transformation due to globalization.

The human resource function has been facing an evolution path from administrative to strategic. Even though one may argue that both need to be balanced and calibrated, based on the premise that, if the administrative work is not done efficiently, accurately, and timely, human resource professionals cannot plan and assume strategic roles. How to balance and/or be dualistic in these roles will remain an ongoing challenge for the profession. From this line of thinking, another question relates to the voice human resources will place in the contemporary context, to be active in facing its dynamics; under what circumstances should human resource professionals become more assertive and take a stand? There is a balance to find, shifting from advocacy and acquiescence, or in other words between a proactive and a reactive attitude and mindset because we face a need to learn when to have a unique point of view which should be articulated and advocated and when to enlist as part of the team/process. We can extend this logic to foresee another debate, shedding light on the core of human resource work and which type of professionals will perform it and how will they be accountable for that. It is important to clearly define roles and accountabilities, once multiple roles will be played by human resource professionals depending on the business context and the proclivity of such professionals.

Rethinking human resource management target audience also calls for attention: a balance shall be found in the quartet workforce vs. customer vs. investor vs. government. As per the one hand, investments in human resource management practices may be focused on improving the workforce (this meaning, ensuring more competent, committed, and dedicated employees), serving external customers (creating organizational capabilities that customers value and pay for), investors (reducing cost, leading to profitability), and/or government (crafting policies with national and/or global impact). On the other hand, we need to keep in mind that each audience has merit, so the myriad of points of view shall be articulated and fulfilled in the best possible way, adopting a holistic approach rather than an exclusive one. We foresee, then, that balancing the needs of multiple audiences raises questions about the focus of human resource management.

This question relates to another, which focuses on the human resource management work scope: shall it be designed as domestic and/or global? Globalization has, undeniably, transformed from a trendy concept to be crystallized into a reality; however, it is an agreed point in the literature that many human resource management processes, practices, and approaches still need rethink so that local organization needs are served along with global requirements.

Therefore, human resource management constituents also call out for attention on this debate; within firm vs. across alliances? A constant trend is pointed in the literature, stating that it will be crucial to learn how to leverage human resource management work both within the firm and across firm alliances. A number of academic research points out that future human resource management work will be housed in organizations that greatly differ from what we have as a standard today. Thus, as organizations become more an assortment of alliances and relationships, so will human resource management practices need to cross boundaries (or to redefine them), in the premise that legal definitions of firm boundaries cannot assume as obstacles for such practices.

This redefinition on processes, practices, and approaches sheds light on the benchmarking debate as a continuous worry on human resource management, specifically in the current context. Following the contribution of institutionalization theory, which reviews the process of sharing ideas across boundaries (crucial in today's world, due to the globalization imprint), in the human resource management world, this shall be operationalized on a constant basis (for example making use of conferences, publications, consultants, and other forums where ideas are quickly shared from one unit to another). Therefore, Ulrich (1997) soon back advised human resource management professionals to be proactive, becoming masters of benchmarking by avoiding the trap "if they did it, so must we."

Current literature on this topic also predicts a more rigorous and demanding measurement process for human resource management in the upcoming times, trying to understand deeper how do we know when human resource management works, as well as how do we tie its work to business results. Therefore, the literature predicts that, in the next few years, we will see more precise, valid, and reliable measures of human resource management effectiveness. This derives from another recurrent issue on this topic, summarized under what Ulrich (1997: 178) labeled "deliverables more than doables." For decades, human resource management has focused mainly on performing good work through the design of programs affecting people and processes. Nevertheless, progressively the emphasis must be shifted to an increasing focus on the deliverables. Deliverables represent the results or outcomes of doing good human resource management work, logic that provides the desired insights and improvements required to face the pace of constant challenge in today's world, as mentioned at the beginning of this essay. So, the logic shall be more ambitious, go deeper, and question what are the organizational implications because innovative staffing, training, or high-performing team programs have been crafted. Undeniably, the literature points a number of possible deliverables, such as turning employees into volunteers, implementing strategy, creating economic value, ensuring cultural heritage, managing employee work/life needs, globalization, among others.

To sum up, instead of predicting a dramatically different future for human resource management, literature thoughtfully demonstrates that much of the past will be found in the future, so we shall learn from the past as well as not to fear change, once it is prone to assume itself as an ongoing governance issue.

5.5 Conclusion

While some answers concerning the future of human resource management—and its path to a green human resource management approach—seem to be known and shared, there is much more unknown; in these cases, the questions are more compelling than the answers. In line with Renwick et al. (2012), we suggest that green human resource management has considerable potential as a research area, so we pose the challenge for further research hoping to reduce the current gap. Nonetheless, with this essay, we will be adding a fruitful contribution in responding to calls in the literature to integrate environmental management and human resource management as a subject of research (Jabbour & Santos, 2008; Jackson et al., 2011), as well as to expand the scope of strategic human resource management to incorporate environmental and sustainability issues (Osland & Osland, 2007; Wilkinson et al., 2001).

We can note that the path to the implementation of a green human resource management paradigm lies on the setup of supportive cultures for environmental management, this based on the fact that effective outcomes in terms of environmental management are achieved not just by making changes to production processes, products, or raw material, instead by deeply adapting the corporate culture so that we can have organizations with profoundly embedded values supportive of long-term sustainability and green management (Kitazawa & Sarkis, 2000; Stone, 2000). Thus, an organizational culture that supports environmental management is one that encourages employees to make suggestions and that ensures them as having the freedom to engage in activities to protect the environment, both at an organizational as well as a personal level. From here, we derive the importance of employee involvement, in line with Fernández et al. (2003), since it forms a core element of an advanced environmental approach because it supports a work culture based on ecological values. Thus, employee involvement in environmental management activities is indispensable for the organization's environmental management efforts to be successful (Fernández et al., 2003).

Therefore, the concepts of *sustainability* and *green management* apply to human resource management, so that a wider green human resource management practice would help place sustainability at the heart of people management, based on the argument that human resources shall be developed and maintained (Ehnert and Harry, 2009). In line with this, we believe that green human resource management has the potential to contribute positively to both employee well-being and improved organizational performance. For the organization, empirical evidence highlights that better environmental performance associates with improved financial performance outcomes, which has been crystallized in the "green pays" argument (Ambec & Lanoie, 2008; Crotty & Rodgers, 2012). For employees, green human resource management practices are likely to improve their well-being in the workplace, improving the working environment and satisfying the needs of an increasingly environmentally aware workforce.

Implementing an effective corporate green management system implies, as core, promoting a great deal of technical and management skills among all employees in the organization (Unnikrishnan & Hedge, 2007), once the development of innovative tools and initiatives of environment management will significantly impact the firm's sustainability and promote a competitive advantage (Hart, 1997). So, the following green human resource processes become vital to foster "green" organizations: strict recruitment and selection strategies based on environmental consciousness (Grolleau et al., 2012), an evaluation process transfiguring into an appraisal and reward systems covering environmental awareness (Jabbour et al., 2010), as well as training and empowerment programs on that topic (Unnikrishnan & Hedge, 2007). For today's organizations, it is crucial to promote a proper alignment of human resource management principles with green management: the greater the strength of green human resource policies, the greater the companies' intensity to adopt environmental management systems and policies, aiming to foster prosperity beyond economic growth (Bohdanowicz, Zientara, & Novotna, 2011).

To conclude, based on the contributions of Dryzek (2005) and Ferrão (2014), we suggest that two main trends of change currently coexist: the one of the *green growth economy* and the other of the *well-being economy*.

Both assume as a basic vector the existence of "clear limits to the exploitation of available resources," advocating the need for a profound change on the pursued aims. Following Vickers (2005) and Jabbour and Santos (2008), we inherently refer to *sustainability* as an imperative in the present and future of human resource management, under the framework of a green human resource management approach. The model presented in Fig. 5.1 (please see above) schematizes this.

References

Ahmad, S. (2015). Green human resource management: Policies and practices. *Cogent Business & Management, 2*(1), 1030–1817.

Ahuja, D. (2015). Green HRM: Management of people through commitment towards environmental sustainability. *International Journal of Research in Finance and Marketing, 5*(7), 50–54.

Ambec, S., & Lanoie, P. (2008). Does it pay to be green? A systematic overview. *The Academy of Management Perspectives, 22*, 45–62.

Ashraf, F., Ashraf, I., & Anam, W. (2015). Green HR for businesses. *International Journal of Academic Research in Business and Social Sciences, 5*(8), 149–156.

Avant, D., Finnemore, M., & Sell, S. (2010). *Who governs the globe?*. Cambridge: Cambridge University Press.

Backhaus, K., Stone, B., & Heiner, K. (2002). Exploringthe relationship between corporate social performance and employer attractiveness. *Business & Society, 41*(3), 292–318.

Band, D., Scanlan, G., & Tustin, C. (1994). Beyond the bottom line: Gain sharing and organizational development. *Personnel Review, 23*(8), 17–32.

Bansal, P., & Roth, K. (2000). Why companies go green: A model of ecological responsiveness. *Academy of Management Journal, 43*, 717–736.

Barney, J. (1991). Firm resources and sustained competitive advantage. *Journal of Management, 17*(1), 99–120.

Bashford, S. (2008). Brownie points for green workers (incentives and motivation). *Human Resource Management International Digest, 16*(7), 30–31.

Basu, K., & Palazzo, G. (2008). Corporate social responsibility: A process model of sensemaking. *Academy of Management Review, 33*(1), 122–136.

Bauer, T., & Aiman-Smith, L. (1996). Green career choices: The influence of ecological stance on recruiting. *Journal of Business and Psychology, 10*(4), 445–458.

Behrend, T., Baker, B., & Thompson, L. (2009). Effects of pro-environmental recruiting messages: The role of organizational reputation. *Journal of Business and Psychology, 24*(3), 341–350.

Berrone, P., & Gomez-Mejia, L. R. (2009). Environmental performance and executive compensation: An integrated agency-institutional perspective. *Academy of Management Journal, 52*(1), 103–126.

Berry, M., & Rondinelli, D. (1998). Proactive corporate environmental management: A new industrial revolution. *Academy of Management Perspectives, 12*(2), 38–50.

Bohdanowicz, P., Zientara, P., & Novotna, E. (2011). International hotel chains and environmental protection: An analysis of Hilton's we care! programme (Europe, 2006–2008). *Journal of Sustainable Tourism, 19*(7), 797–816.

Boiral, O. (2002). Tacit knowledge and environmental management. *Long Range Planning, 5*(3), 291–317.

Branzei, O., Ursacki-Bryant, T., Vertinsky, I., & Zhang, W. (2004). The formation of green strategies in Chinese firms: Matching corporate environmental responses and individual principles. *Strategic Management Journal, 25*(11), 1075–1095.

Brekke, K., & Nyborg, K. (2008). Attracting responsible employees: Green production as labor market screening. *Resource and Energy Economics*, *30*(4), 509–526.
Brockett, J. (2006). Change agents. *People Management, November*, 18–19.
Cassio, J., & Rush, A. (2009). *Green careers: Choosing work for a sustainable future*. Gabriola Island: New Society Publishers.
Cherian, J., & Jacob, J. (2012). A study of green HR practices and its effective implementation in the organization: A review. *International Journal of Business and Management*, 7(21), 25–33.
Chiang, C. (2010). Perceived organizational change in the hotel industry: An implication of change schema. *International Journal of Hospitality Management*, *29*(1), 157–167.
Chinander, K. (2001). Aligning accountability and awareness for environmental performance in operations. *Production and Operations Management*, *10*(3), 276–291.
Christmann, P. (2000). Effects of "best practices" of environmental management on cost advantage: The role of complementary assets. *Academy of Management Journal*, *43*(4), 663–680.
Coff, R., Coff, D., & Eastvold, R. (2006). The knowledge-leveraging paradox: How to achieve scale without making knowledge imitable. *Academy of Management Review*, *31*(2), 452–465.
Collins, J., & Hitt, M. A. (2006). Leveraging tacit knowledge in alliances: The importance of using relational capabilities to build and leverage relational capital. *Journal of Engineering and Technology Management*, *23*(3), 147–167.
Crotty, J., & Rodgers, P. (2012). Sustainable development in the Russia federation: The limits of greening within industrial firms. *Corporate Social Responsibility and Environmental Management*, *19*(3), 178–190.
Daily, B., Bishop, J., & Govindarajulu, N. (2009). A conceptual model for organizational citizenship behavior directed toward the environment. *Business & Society*, *48*(2), 243–256.
Daily, B., Bishop, J., & Steiner, R. (2007). The mediating role of EMS teamwork as it pertains to HR factors and perceived environmental performance. *Journal of Applied Business Research*, *23*(1), 95–109.
Daily, B., & Huang, S. (2001). Achieving sustainability through attention to human resource factors in environmental management. *International Journal of Operations & Production Management*, *21*(12), 1539–1552.
Darnall, N., & Edwards, D., Jr. (2006). Predicting the cost of environmental management system adoption: The role of capabilities, resources and ownership structure. *Strategic Management Journal*, *27*(4), 301–320.
Datta, M. (2015). Green work-life balance: A new concept in green HRM. *International Journal of Multidisciplinary Approach & Studies*, *2*(2), 83–89.
Dechant, K., & Altman, B. (1994). Environmental leadership: From compliance to competitive advantage. *Academy of Management Executive*, *8*(3), 7–27.
del Brío, J., Fernández, E., & Junquera, B. (2007). Management and employee involvement in achieving an environmental action-based competitive advantage: An empirical study. *The International Journal of Human Resource Management*, *18*(4), 491–522.
Denton, D. (1999). Employee involvement, pollution, control and pieces to the puzzle. *Environmental Management and Health*, *10*(2), 105–111.
Dryzek, J. (2005). *The politics of the earth: Environmental discourses*. Oxford: Oxford University Press.
Egri, C., & Herman, S. (2000). Leadership in the North American environmental sector: Values, leadership styles, and contexts of environmental leaders and their organizations. *Academy of Management Journal*, *43*(4), 571–604.
Ehnert, I., & Harry, W. (2012). Recent developments and future prospects on sustainable human resource management: Introduction to the special issue. *Management Revue*, *23*(3), 221–238.

Epstein, M., & Roy, M. (1998). Managing corporate environmental performance: A multinational perspective. *European Management Journal, 16*(3), 284–296.

Fernández, E., Junquera, B., & Ordiz, M. (2003). Organizational culture and human resources in the environmental issue: A review of the literature. *International Journal of Human Resource Management, 14*(4), 634–656.

Ferrão, J. (2014). (Coord.). *Que Economia Queremos?*. Lisboa: Fundação Francisco Manuel dos Santos.

Florida, R., & Davison, D. (2001). Gaining from green management: Environmental management systems inside and outside the factory. *California Management Review, 43*(3), 64–84.

Forman, M., & Jorgensen, M. (2001). The social shaping of the participation of employees in environmental work within enterprises—Experiences from a Danish context. *Technology Analysis & Strategic Management, 13*(1), 71–90.

Fryxell, G., & Lo, C. (2003). The influence of environmental knowledge and values on managerial behaviours on behalf of the environment: An empirical examination of managers in China. *Journal of Business Ethics, 46*(1), 45–69.

Govindarajulu, N., & Daily, B. (2004). Motivating employees for environmental improvement. *Industrial Management & Data Systems, 104*(4), 364–372.

Grolleau, G., Mzoughi, N., & Pekovic, S. (2012). Green not (only) for profit: An empirical examination of the effect of environmental-related standards on employees' recruitment. *Resource and Energy Economics, 34*(1), 74–92.

Haden, S., Oyler, J., & Humphrey, J. (2009). Historical, practical, and theoretical perspectives on green management. An exploratory analysis. *Management Decision, 47*(7), 1041–1055.

Hart, S. (1997). Beyond greening: Strategies for a sustainable world. *Harvard Business Review, 75*(1), 66–77.

Henriques, I., & Sadorsky, P. (1996). The determinants of an environmentally responsive firm: An empirical approach. *Journal of Environmental Economics and Management, 30*(3), 381–395.

Jabbour, C., & Santos, F. (2008). The central role of human resource management in the search for sustainable organizations. *The International Journal of Human Resource Management, 19*(12), 2133–2154.

Jabbour, C., Santos, F., & Nagano, M. (2010). Contributions of HRM throughout the stages of environmental management: Methodological triangulation applied to companies in Brazil. *The International Journal of Human Resource Management, 21*(7), 1049–1089.

Jackson, S., Renwick, D., Jabbour, C., & Muller-Camen, M. (2011). State-of-the-art and future directions for green human resource management: Introduction to the special issue. *German Journal of Human Resource Management, 25*(2), 99–116.

Ji, L., Huang, J., Liu, Z., Zhu, H., & Cai, Z. (2012). The effects of employee training on the relationship between environmental attitude and firms' performance in sustainable development. *The International Journal of Human Resource Management, 23*(14), 2995–3008.

Khalili, N., & Duecker, S. (2013). Application of multi-criteria decision analysis in design of sustainable environmental management system framework. *Journal of Cleaner Production, 47*, 188–198.

Kitazawa, S., & Sarkis, J. (2000). The relationship between ISO 14001and continuous source reduction programs. *International Journal of Operations and Production Management, 20*(2), 225–248.

Kramar, R. (2014). Beyond strategic human resource management: Is sustainable human resource management the next approach? *The International Journal of Human Resource Management, 25*(8), 1069–1089.

Kunes, T. (2001). A green and lean workplace? *Strategic Finance, 82*(8), 70–73.

Laabs, J. (1992). The greening of HR. *The Personnel Journal, 71*(8), 60–68.

Lai, K., Cheng, T., & Tang, A. (2010). Green retailing: Factors for success. *California Management Review, 52*(2), 6–31.

Lenssen, G., Tyson, S., Pickard, S., Bevan, D., Castelló, I., & Lozano, J. (2009). From risk management to citizenship corporate social responsibility: Analysis of strategic drivers of change. *Corporate Governance: The International Journal of Business in Society, 9*(4), 373–385.

Llewellyn, B., & Golden, K. (2008). *Green jobs: A guide to eco-friendly employment*. Avon: Adams.

Lorenzi, N., & Riley, R. (2003). Organizational issues = change. *International Journal of Medical Informatics, 69*(2–3), 197–203.

Mandip, G. (2012). Green HRM: People management commitment to environmental sustainability. *Research Journal of Recent Sciences, 1*, 244–255.

Massoud, J., Daily, B., & Bishop, J. (2008). Reward for environmental performance: Using the Scanlon plan as catalyst to green organisations. *International Journal on Environment, Workplace and Employment, 4*(1), 15–31.

May, D., & Flannery, B. L. (1995). Cutting waste with employee involvement teams. *Business Horizons, 38*(5), 28–39.

McDonagh, P., & Prothero, A. (1997). *Green management: A reader*. London: Dryden Press.

Milliman, J., & Clair, J. (1996). Best environmental HRM practices in the U.S. In W. Wehrmeyer (Ed.), *Greening people: Human resources and environmental management* (pp. 49–73). Sheffield: Greenleaf Publishing.

Muster, V., & Schrader, U. (2011). Green work-life balance: A new perspective for green HRM. *German Journal of Human Resource Management, 25*(2), 140–156.

Opatha, H., & Arulrajah, A. (2014). Green human resource management: Simplified general reflections. *International Business Research*, 7(8), 101–112.

Osland, A., & Osland, J. (2007). Aracruz Celulose: Best practices icon but still at risk. *International Journal of Manpower, 28*(5), 435–450.

Owen, D. (Ed.). (1992). *Green reporting: Accountancy and the challenge of the nineties*. London: Chapman Hall.

Parks, B., & Helmer, J. (2009). *Green careers*. New York: Alpha Publishing.

Peacock, M. (1993). Developing environmental performance measures (ways to measure the performance of corporate environmental programs). *Industrial Engineering, 25*(9), 205–223.

Peattie, K. (1992). *Green marketing*. London: Pitman.

Perron, G., Côté, R., & Duffy, J. (2006). Improving environmental awareness training in business. *Journal of Cleaner Production, 14*(6–7), 551–562.

Ramus, C. (2001). Organizational support for employees: Encouraging creative ideas for environmental sustainability. *California Management Review, 43*(3), 85–105.

Ramus, C., & Steger, U. (2000). The roles of supervisory support behaviours and environmental policy in employee "ecoinitiatives" at leading-edge European companies. *Academy of Management Journal, 43*(4), 605–626.

Rees, S. (1996). Action through ownership: Learning the way at Kent County council. In W. Wehrmeyer (Ed.), *Greening people: Human resources and environmental management* (pp. 357–375). Sheffield: Greenleaf Publishing.

Remmen, A., & Lorentzen, B. (2000). Employee participation and cleaner technology: Learning processes in environmental teams. *Journal of Cleaner Production, 8*(5), 365–373.

Renwick, D., Redman, T., & Maguire, S. (2012). Green human resource management: A review and research agenda. *International Journal of Management Reviews, 15*(1), 1–14.

Reynolds, R., & Ablett, A. (1998). Transforming the rhetoric of organizational learning to the reality of the learning organization. *The Learning Organization, 5*(1), 24–35.

Risher, H. (2003). Refocusing performance management for high performance. *Compensation and Benefits Review, 35*(5), 20–30.

Roome, N. (1992). Developing environmental management strategies. *Business Strategy and the Environment, 1*(1), 11–24.

Rothenberg, S. (2003). Knowledge content and worker participation in environmental management at NUMMI. *Journal of Management Studies, 40*(7), 1783–1802.

Roy, M., & Thérin, F. (2008). Knowledge acquisition and environmental commitment in SMEs. *Corporate Social Responsibility and Environmental Management, 15*(5), 249–259.

Russo, M., & Fouts, P. (1997). A resource-based perspective on corporate environmental performance and profitability. *Academy of Management Journal, 40*(3), 534–559.

Russo, M., & Harrison, N. (2005). Organizational design and environmental performance: Clues from the electronics industry. *Academy of Management Journal, 48*(4), 582–593.

Siebenhüner, B., & Arnold, M. (2007). Organizational learning to manage sustainable development. *Business Strategy and the Environment, 16*(5), 339–353.

Simms, J. (2007). *Direct action* (pp. 36–39). People Management. 26th July.

Stanwick, P., & Stanwick, S. (2001). CEO compensation: Does it pay to be green? *Business Strategy and the Environment, 10*(3), 176–182.

Stone, L. (2000). When case studies are not enough: The influence of corporate culture and employee attitudes on the success of cleaner production initiatives. *Journal of Cleaner Production, 8*(5), 353–359.

Tang, G., Chen, Y., Jiang, Y., Paille, P., & Jia, J. (2018). Green human resource management practices: Scale development and validity. *Asia Pacific Journal of Human Resources, 56*(1), 31–55.

Tariq, S., Jan, F., & Ahmad, M. (2016). Green employee empowerment: A systematic literature review on state-of-art in green human resource management. *Quality & Quantity, 50*(1), 237–269.

Taylor, S. (1992). Green management: The next competitive weapon. *Futures, 24*(7), 669–680.

Teece, D. (1977). Technology transfer by multinational firms: The resource cost of transferring technological know-how. *The Economic Journal, 87*(346), 242–261.

Tsai, W. (2001). Knowledge transfer in intraorganizational networks: Effects of network position and absorptive capacity on business unit innovation and performance. *Academy of Management Journal, 44*(5), 996–1005.

Ulrich, D. (1997). HR of the future: Conclusions and observations. *Human Resource Management, 36*(1), 175–179.

Unnikrishnan, D., & Hedge, S. (2007). Environmental training and cleaner production in Indian industry–a micro-level study. *Resources Conservation and Recycling, 50*(4), 427–441.

Vickers, M. R. (2005). Business ethics and the HR role: Past, present, and future. *Human Resource Planning, 28*(1), 26–32.

Vlados, C., Deniozos, N., & Chatzinikolaou, D. (2018). Global crisis, innovation and change management: Towards a new systemic perception of the current globalization restructuring. *International Business Research, 11*(8), 9–29.

Waddock, S. (2000). The multiple bottom lines of corporate citizenship: Social investing, reputation, and responsibility audits. *Business and Society Review, 105*(3), 323–345.

Wagner, M. (2013). 'Green' human resource benefits: Do they matter as determinants of environmental management system implementation? *Journal of Business Ethics, 114*(3), 443–456.

Walton, S., Handfield, R., & Melnyk, S. (1998). The green supply chain: Integrating suppliers into environmental management processes. *International Journal of Purchasing and Materials Management, 34*(1), 2–11.

Wehrmeyer, W. (Ed.). (1996). *Greening people: Human resources and environmental management.* New York: Routledge.

Wilkinson, A., Hill, M., & Gollan, P. (2001). The sustainability debate. *International Journal of Operations and Production Management, 21*(12), 1492–1502.

Yeung, A., Woolcock, P., & Sullivan, J. (1996). Identifying and developing HR competencies for the future. *Human Resource Planning, 19*(4), 48–58.

Young, W., & Welford, R. J. (1998). An environmental performance measurement framework for business. *Greener Management International, Spring*(21), 30–50.

CHAPTER 6

Factors involved in using plastic money in the Jangalmahal region of West Bengal: An empirical study

Sudin Bag[a] and Nilanjan Ray[b]
[a]Department of Business Administration, Vidyasagar University, Midnapore, West Bengal, India
[b]Marketing and Examination Coordinator, Adamas University, Kolkata, India

6.1 Introduction

In light of globalization, we reach the global frontier from all aspects. We are changing our food habits, culture, taste, and others, i.e., from conventional to continental, Indian to Western, and local to global. The financial sectors are not except for this reflection of globalization light. In the line of advancement of technology in the monetary arrangements, India has come forward over time from the barter (ancient) system to the more convoluted forms, i.e., coins, cash, and cheques. In the 21st century, payment through cash and cheques itself has encountered a transfiguration. It has moved from being a paper-based transfer of value to virtual electronic one. Electronic service is becoming a feasible option for dealings between financial service providers and their clients. With the advent of technology, an impressive reallocation in the financial sector has been observed primarily in the banking sector in India. One of such remarkable additions in the banking sector is "plastic money."The growth and development of plastic money are doubtlessly one of the most remarkable phenomena in six of the modern banking era, making a significant impact on the financial sector. The utilization of plastic money has appeared as a type of amenity to the financial institution customers. Electronic service is becoming a feasible option for dealings between financial service providers and their clients (Rotchanakitumnuai & Speece, 2004).

In recent times, uses of plastic money, namely debit card, credit card, cash card, has come as a form of amenity to financial institution customers. This electronic form of payment has far-reaching significance notably for India. Primarily, it helps to align with the global financial service businesses by

Green Production Engineering and Management
https://doi.org/10.1016/B978-0-12-821238-7.00004-X

making the systems more competent and cost-effective. It also supports in undertaking the unique challenges India faces the large unbanked society, deregulated cash economy, and need for monetary clarity. Plastic money yields many more benefits than merely facilitating the replacement of cash and cheques for consumer transactions. It has an extensive brunt on the financial system as a whole through low transaction costs, increased operational competence, improved financial framework, and an opportunity for the expeditiously evolving Indian economy to globalize further. These debit and credit cards are gradually replacing the traditional system of payment through cash. Banks are providing facilities regarding plastic money for competing to attract customers to use plastic money.

Preference for using plastic money by the customers is proliferating. In recent times, people are not willing to carry cash with them; instead of that, they like to hold cards in their purses or wallets.

Apart from retreating cash from an ATM or transacting at a POS (Point of Sale) in a retail market, plastic money conveys its improved security, portability, 24-h user-friendliness to access, secure payment of monthly bills, and transfer funds between the accounts in minimal time. In a developing country like India, higher value purchases are increasingly being made through plastic money, and customers get the opportunity to buy high-value products in installments.

Plastic money is progressively becoming the principal mode of transaction in India due to some immense facilities. However, a few problems triumph concerning using plastic money, especially in the context of rural India. So, if the issues and the projection that are faced by the crucial users be known, that can be investigated, and empirical study can be taken consequently to improve. In that association, the study is an endeavor to unveil the effectiveness of plastic money in India with particular emphasis on the Jangalmahal region of West Bengal.

The communal effectiveness index (CEI) has been developed to highlight the significance of the community effectiveness on plastic money. The CEI is a relatively simple indicator that divides total plastic money users on the communal effectiveness by the total number of respondents of this study. For example, if the CEI were 0.8, this index number would mean that the community effectively represents 80% of the total communal effectiveness of plastic money acceptance. The higher index showcases the better community's adoption toward plastic money. The example of a CEI of 0.861 can also assist to explain how the CEI assesses the significance of the major factors of communal support, communal security, and communal service. For instance, the CEI for communal support might be 0.888; the CEI for communal security, 0.850; and the CEI for communal service purpose, 0.844. These three program-level CEIs necessarily add up to the overall CEI {e.g., (0.888 + 0.850 + 0.844)/3 = 0.861}. This example is similar to the actual result across the Jangalmahal. Three factors of communal effectiveness are indeed the principal form of communal effectiveness.

6.2 Literature review

Mester (2006) reported that the usage of debit cards had risen from about 18% of households in 1995 to 59% in 2004. This growing trend is universal to all income, sex, and age groups. Another feature is that the rich use debit cards much more frequently than the poor people. In 2004, around 41% of low-income households used debit cards, while that of upper-income households reached almost 70%.

Hafalir and Loewenstein (2012) in their article, persistent on two categories of customers, one is revolvers (who carry debt) and another is convenience users (who do not carry debt), measured the impact of payment with a credit card as compared to cash by an insurance company employee's spending on lunch in a cafeteria. It was found that there was a change in the dinner's payment method from the cash to a credit card when an inducement to pay with a credit card was certain. It was then found out that credit cards do not increase expenditure. However, the utilization of credit cards has a degree of different impacts on spending for convenience and revolver users. Revolvers (who carry debt) spend less when induced to pay with a credit card, whereas convenience users display the opposite pattern.

However, Manivannam (2013) pointed out that the prospects and usability of plastic money are not the same everywhere. It was observed that the young adults seek fulfillment through hedonic activities such as shopping, hanging out, and great lifestyles by using credit cards (Hausman, 2000).

In another study, Soman and Cheema (2002) investigated that intensity to amplify as a purpose of the credit limit, particularly as credit limit increases, subject to the utilization of credit card accounted a greater probability of making a purchase, other things remaining constant. When a consumer uses a credit card and a credit line available, sometimes unnecessary spending gets unavoidable. Sometimes, gender and educational background also play a role in holding and usages of plastic money (Kaseke, 2012).

Bisht, Nair, Dubey, and Hajela (2015) in their study showcased that the consumers usually do not embrace adequate cash to buy anything; instead, they can utilize their cards to purchase. Debit or credit cards (plastic money) are sporadically provided benefits for the users, and sometimes, it bothers due to transactional payments and unnecessary official procedures.

According to Feinberg (1986), people used plastic cards in the form of plastic money, as a sort of mental satisfaction might take place, by which card impetus may be related to spending. Four experiments were conducted to identify the relationship between situational conditioning and consumer behavior toward the usage of a credit card as a payment mode. These experiments resulted that plastic money, especially credit card, impetus can increase the immensity, probability, and time taken to make decisions associated with spending (Patel & Amin, 2012).

Bernthal, Crockett, and Rose (2005) pointed out that credit cards are proficient in moving customers toward a specific lifestyle. The usage pattern of plastic money depicts particular values and users' lifestyle patterns. The usage pattern of plastic cards is dependent on holders "Internalization of ideologies of entitlement and frugality." Apart from this, demographic and psychographic variables are playing an important role in the marketing of plastic cards. The perception of Indian consumers toward the usage of the plastic card might relate to satisfaction and a sense of gratification. Although "status" is not associated with the use of plastic money to the Indian card users (Khare, Khare, & Singh, 2012), but for Chinese card users, status is an important issue (Worthington, Stewart, & Lu, 2007) and for Australian customers too (Phau & Woo, 2008). However, Barker, D'Amato, and Sheridon (2008) pointed out that the prosperous Chinese consumers in urban areas already use plastic money for making payment of high-value purchase and such kinds of users are mostly the innovators of these products, whereas Lee and Huang (2011) reported that the plastic money, especially credit cards, is tapping the mass market @ 204.3% in Taiwan. The major dominant factors that are associated with the rapid growth of card markets are *per capita* income, savings, disposable income, and the total count of banks in Taiwan. The consumers in Malaysia came out as extensive spenders, and most of them go on spending aggressively even at the time of soaring prices. It found some interrelation between influencing components likewise, demographic ingredients, banking policies, and consumer inclination toward money, and consumption behavior through plastic money in Malaysia (Ming-Yen Teoh, Chong, & Mid Yong, 2013). In line with this, Nga, Yong, and Sellappan (2011) stated that consumer behaviors such as awareness level, materialism, and compelled spending have a positive impact on the usage behavior of plastic cards among Malaysian students. In a study in Korea, female card users also were explaining the effect of plastic cards on fashion orientation, fashion inducement, and compulsive buying (Park & Davis, 2005).

On the contrary, Brito and Hartley (1995) argued that card users prefer to pay the interest amount on the due balance on credit cards rather than spending the negotiation cost related to the scheduling of loans from the banking institutions in spite of heavy interest charges. A person's expectations and personal characteristics have an impact on the attitude toward the payment method. In addition to this, Chin, Yi, Kwan, and Si (2018) shed light on the ways that generally the family conversation and interaction between parents and young consumers result in a discussion of credit card usage behavior. Chakravorti and Emmons (2001) reported that the reputation of plastic money as a payment medium has been associated with the facilitation of not carrying cash and cheques, the limited liability of lost and stolen cards, and additional augmentations, such as dispute resolution services and perks.

Dhanda and Arora (2017) suggested that the use of plastic money in India is booming attributable to an increase in income and revolutionary growth in the technology and retail sectors. Young Indians feel dignified in holding plastic money, and they think about it as an excellent approach to paying. Although the utilization of plastic cards in India is without any doubt on the rise from the previous few years, however, there is still a good potential left for the bankers to introduce a lot of enticing services to tempt the shoppers on one facet and improve their gain on the opposite.

Annamalai and Muthu (2008) projected the growth of debit and credit cards in the retail transactions. They also mentioned the growth factors, which lead to its popularity, significant constraints faced by banks and summarized with bright future and scope of plastic money.

Patel & Amin, 2012 suggested that due to the technology, changes bring the new shape in the world economy. And among all, the changes in the economy lead to some drastic changes in the transaction. Nowadays, in any transaction, plastic money becomes an inevitable part of the transaction. And with it, life becomes more comfortable, and development would take a better place. Relating to the Indian scenario, how plastic money took place in the banking world would be the focus of the researcher over here. And along with the plastic money, it becomes possible that control the money laundry and effective utilization of financial system would become possible which would also help for tax legislation. The significant factors that influence the users toward plastic money are service offers, promotional offers, interest benefits, cash benefits, ease of payments, payment charges, card benefits, and time benefits (Nayak & Agarwal, 2008).

6.3 Objectives of the study

The primary aim of this study was to examine the effectiveness of plastic money as an alternative to cash in India with special reference to Jangalmahal, West Bengal.

Along with the common objective, some specific objectives will also be examined. Those are as follows:

- To identify the demographic profile of the users,
- To understand the influencing factors behind the use of plastic money,
- To measure the effectiveness of the use of plastic money in the Jangalmahal region.

6.4 Research methodology

The article is based on an empirical study of factors that influencing the users of plastic money and usage patterns of plastic money. The article takes into consideration the positive as well as negative aspects of using plastic money over real money.

The primary aim of this research paper was to identify the actual scenario of plastic money in rural India with special reference to Jangalmahal. In this study, sources of data collection have been considered like secondary and primary. As part of secondary sources, data have been drawn from the literature review, focused text, administrative files, and even some institutional files. The primary data have been gathered through a field survey using an undisguised structured questionnaire. The questionnaire has been designed keeping in view the general uses and problems that a consumer faces when using plastic money. Demographic factors such as age, income level, gender, and profession have also been considered. The questionnaire was based on a 5-point Likert scale. The 5-point Likert scale chosen for the study has labels attached to each point on the mentioned scale, and the midpoint of the scale has been set as the point of neutrality. The label is as follows: 1: strongly disagree; 2: disagree; 3: neutral; 4: agree; 5: strongly agree.

The jurisdiction of the survey encompasses three districts of West Bengal, namely West Midnapore, Purulia, and Bankura, which are alternatively known as Jangalmahal in West Bengal.

The questionnaire was tested through a pilot study on 40 consumers. The pilot study helped in making the needed changes in the questionnaire and allowed the researchers to have a more clear knowledge of what consumers' perception is toward the usage of plastic money, thereby leading to an improved questionnaire. The refined questionnaire has been asked to a total of 205 respondents from the entire jurisdiction of the study, and SPSS 23 has been used to analyze the responses of the survey.

The CEI's construction is quite simple. It is based on comparing two ratios. The total communal effectiveness index is the average of the factors on communal security (CS1), communal support (CS2), and communal service (CS3). Similarly, the total number of respondents is the sum of the proposed beneficiaries of all three factors. To simplify, "E" can be used to stand for effectiveness. The total number of the beneficiary is used to calculate the average of the factors, where the total number of users is divided by the total number of an intended recipient. Thus, the CEI formula is as follows:

$$\text{CEI} = \left(\text{ECS1} + \text{ECS2} + \text{ECS3}\right) / 3)$$

$$\text{ECS1} = \frac{\text{Average of the factors}}{\text{Total no of factors}}$$

The implicit weight for each of the three major programs is the number of its intended beneficiaries as a ratio of the average of the factors to the total intended recipients of all three major programs. The vital point to keep in mind when this index presents results is that the overall CEI for the community is, in effect, a weighted average of the CEIs for each of the three major programs (communal security, communal support, and community service). The same kind of population weights can be applied to the subcomponents of the three major programs. For example, if time-saving, reduce risk, portability, and easy purchase were the only subcomponents of communal support, the average of their own individual CEIs would be equal to the CEI of communal support as a whole. So, the CEIs of individual subcomponents would always be average to the CEI of the overall program, and the CEIs of the three major programs would be still an average percentage of the overall CEI of the community.

6.5 Analysis and findings

From Table 6.1, it has seen that 84% of the respondents are male and 41% of the respondents belong to the age group between 21 and 30-year age bracket, i.e., the young Indians have been represented in this study. The 27.4% of the respondents are having postgraduation qualification, followed by 25.8% higher secondary and 25.3% graduation degree. As per the occupation, it is concerned 29.3% of respondents are students followed by 27.3%

Table 6.1 Demographic profile of the respondents.

Sl. No.	Factor	Category	No. of respondents	Percentage
1	Gender	Male	173	84.0
		Female	32	16.0
		Total	**205**	**100**
2	Age	Up to 20 years	16	7.8
		21–30 years	84	41.0
		31–40 years	40	19.5
		41–50 years	48	23.4
		Above 50 years	17	8.3
		Total	**205**	**100**
3	Educational qualification	Up to secondary	20	9.8
		Higher secondary	53	25.8
		Graduation	52	25.3
		Postgraduation	56	27.4
		Others	24	11.7
		Total	**205**	**100**
4	Occupation	Agriculture	41	20.0
		Business	56	27.3
		Service	40	19.5
		Students	60	29.3
		Others	8	3.9
		Total	**205**	**100**
5	Monthly income	Up to 5000	40	19.5
		5001–10,000	80	39.0
		10,001–15,000	33	16.1
		15,001–20,000	24	11.7
		Above 20,000	28	13.7
		Total	**205**	**100**

Source: Own Data.

of respondents are involved in the business. In the income pattern of the respondents, it is found that most of the respondents 39% having monthly income Rs.5001–Rs.10000 followed by 19.5% of the respondents monthly income up to Rs. 5000.

From Table 6.2 and Fig. 6.1, it is found that most of the respondents (82%) are holding debit cards as plastic money and only 18% are using debit and credit cards as plastic money to meet their requirements. But, it is quite surprising that no one holds only credit cards as plastic money in the Jangalmahal region.

From Table 3 and Fig. 2, one found that a good number of respondents are using plastic money for shopping followed by cash withdrawal from

Table 6.2 Cardholding pattern of the respondents.

Card holds by the respondents	No. of respondents
Debit	169
Credit	0
Both	36

Source: Own Data.

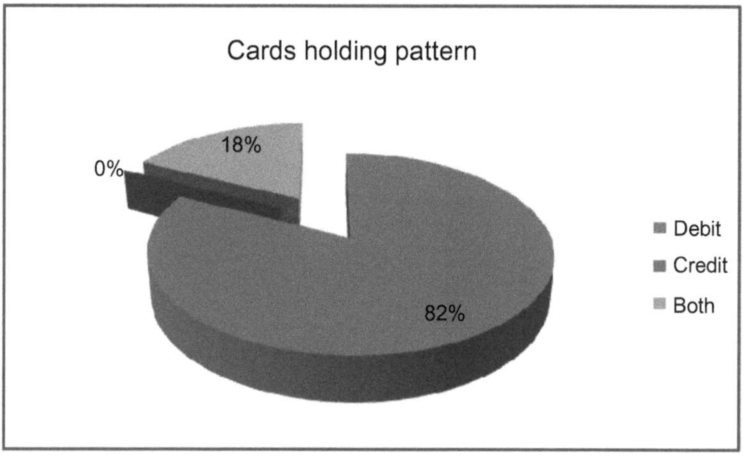

Fig. 6.1 Cardholding pattern of the respondents.

Table 6.3 Distribution of usage of plastic money of the respondents.

usage of plastic money	No of respondents
Shopping	148
ATM	36
Hotel and restaurant	20
Hospital	0
Travel	1

Source: Own Data.

the ATM. It is notable that most of the respondents use cards for online shopping and students are using the plastic money for hotel and restaurant bill payments. It has also been found that the expenditure pattern of the Jangalmahal region has changed by the use of plastic money.

Table 6.4 and Fig. 3 represent the frequency of the use of plastic money. It is found that only four users are using plastic money daily basis. But a good number of respondents using plastic money 2–5 times monthly followed by 6–10 times monthly users. However, the rare users expect that

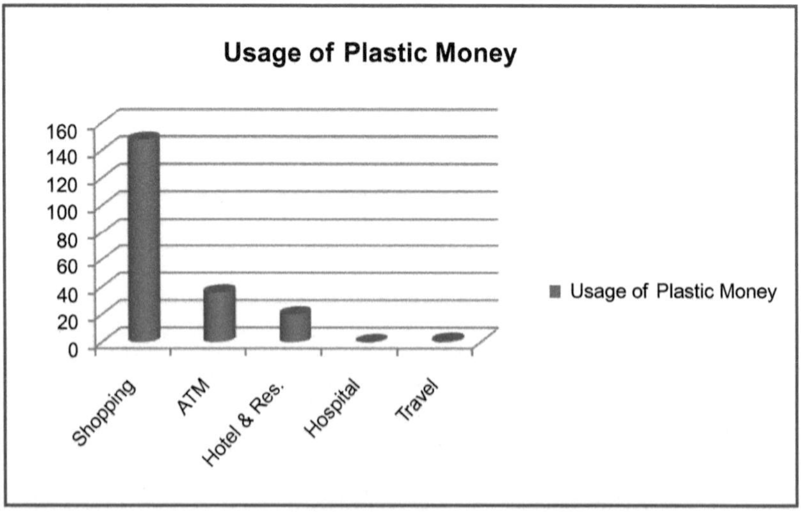

Fig. 6.2 Usages of plastic money.

Table 6.4 Frequency of usage of plastic money.

The frequency of usage of plastic money	No. of respondents
Rarely	58
2–5 times	61
6–10 times	53
11–15 times	29
Daily	4

Source: Own Data.

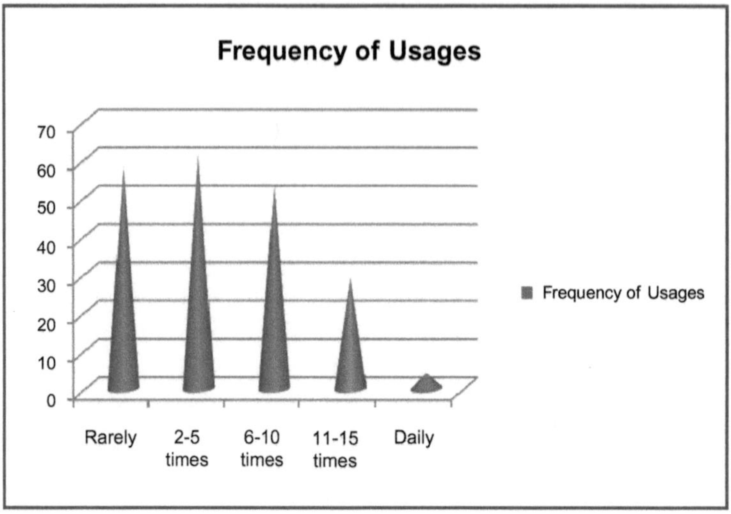

Fig. 6.3 Frequency of uses of plastic money.

they will become regular users from time to time with the effectiveness of the plastic money.

In connection with the identification of the perception of the respondents were asked 15 close-ended questions about the plastic money as an alternative to cash. Responses have been recorded in a 5-point Likert scale from strongly disagree to agree strongly. The descriptive statistics are presented in Table 5.

The result of the descriptive statistics reflects that zero processing fee, security, technological advances, low risk, and addition of value are the essential reasons for adopting and using plastic money in the Jangalmahal region.

The researcher has performed a reliability analysis to check the internal consistency, i.e., the reliability of the questionnaire. The output of the reliability test is presented in Table 6.6.

Table 6.6 represents the reliability statistics and the value of Cronbach's alpha (0.905). It indicates the high internal consistency of the scale (Bag, Aich, & Islam, 2020). Thus, to identify the influencing factors of the perception of the users of a plastic card, the researcher has performed a factor analysis. The result of the factor analysis is presented in Table 6.7.

Table 6.5 Descriptive statistics.

	N	Mean	Std. deviation
Acceptability	205	4.37	0.792
Added value	205	2.90	1.192
Convenience	205	4.41	0.692
Easy purchase	205	4.43	0.935
Extreme benefits	205	3.34	0.965
High accessibility	205	4.14	0.768
Instant transaction	205	4.16	0.801
Low interest	205	3.91	0.998
portability	205	4.30	0.848
Prestige	205	3.82	0.879
Reduce risk	205	4.20	1.221
Secure	205	4.41	1.106
Technology	205	4.21	1.072
Time-saving	205	4.33	0.901
Zero processing fee	205	4.24	1.079
Valid N (listwise)	205		

Table 6.6 Reliability statistics.

Cronbach's alpha	No. of items
0.905	15

Table 6.7 KMO and Bartlett's tests.

Kaiser-Meyer-Olkin measure of sampling adequacy		0.797
Bartlett's test of sphericity	Approx. chi-square	2143.118
	Df	105
	Sig.	0.000

Table 6.7 represents the KMO and Bartlett's tests that indicate the suitability of data for structure detection. The output depicts that KMO and Bartlett's tests satisfy the criteria to perform the factor analysis. The value of KMO is 0.797, which is higher than the threshold value of 0.05 (Bag & Omrane, 2020; Malhotra & Dash, 2010). The Bartlett's test of sphericity (Bartlett, 1950) is another requirement for verifying and employing the factor analysis, and the result in Table 6.7 represents that the test value is 2143.118, which is significant ($P<0.001$). Therefore, data are appropriate to perform the factor analysis.

From Table 6.8, it is found that only three factors have been extracted, i.e., eigenvalue more than 1, and these three factors are explaining almost 67% of the total variables. It is also revealed that the first factor accounts for 46.041, second factor accounts for 13,391, and the third factor accounts for 7.561. All the remaining factors are not significant.

Table 6.9 represents the rotated component matrix, where three suppressed factors are elucidating sufficient variation of the observed variables. The environment of the greater correlated observed variables with the extracted factors; it is being concluded that these extracted latent factors can be named F1 as an influential factor, F2 as security factor, and F3 as cost factor regarding the perception of using plastic money users in Jangalmahal.

The study investigates the relative importance of the main subcomponents of the three major factors (communal security, communal support, and communal service), comparing the number of respondents of each subcomponent.

In **communal support**, time-saving on transfers, reduce risk, portability, and easy purchase each account for about communal support among all of the beneficiaries, where time-saving, portability, and easy purchases have largely determined the character of communal support in Jangalmahal. Among those, the time-saving factor is the most critical subcomponent, accounting for 0.92 of all beneficiaries of this significant program.

In **communal security**, high accessibility, technology, and prestige are the three most essential subcomponents: high accessibility, with 92% average value, 82% of technology, and 80% of prestige beneficiaries in this major

Table 6.8 Total variance explained.

	Initial eigenvalues			Extraction sums of squared loadings			Rotation sums of squared loadings		
Component	Total	% of variance	Cumulative %	Total	% of variance	Cumulative %	Total	% of variance	Cumulative %
1	6.906	46.041	46.041	6.906	46.041	46.041	4.748	31.653	31.653
2	2.009	13.391	59.432	2.009	13.391	59.432	3.334	22.225	53.878
3	1.134	7.561	66.993	1.134	7.561	66.993	1.967	13.115	66.993
4	0.856	5.706	72.699						
5	0.784	5.226	77.925						
6	0.770	5.130	83.055						
7	0.538	3.584	86.639						
8	0.498	3.318	89.957						
9	0.365	2.436	92.392						
10	0.324	2.163	94.555						
11	0.246	1.638	96.193						
12	0.225	1.497	97.691						
13	0.137	0.914	98.605						
14	0.130	0.866	99.471						
15	0.079	0.529	100.000						

Extraction method: principal component analysis.

Table 6.9 Rotated component matrix.[a]

	Component		
	F1	F2	F3
Instant transaction	0.834		
Portability	0.800		
Timesaving	0.774		
Acceptability	0.745		
Easy purchase	0.618		
Reduce risk	0.585		
Added value	0.575		
Convenience	0.547		
Extreme benefits	0.510		
Secure		0.837	
Prestige		0.752	
Technology		0.741	
High accessibility		0.633	
low-interest			0.953
Zero processing fee			0.951

Extraction method: principal component analysis.
Rotation Method:Varimax with Kaiser normalization.
[a] Rotation converged in five iterations.

program. The CEI project groups other communal security programs, such as benefits from high accessibility, technology, and prestige, under the broad category of "other forms of communal security." Another group is communal service programs, such as low interest and zero processing fee. These three programs are together accounted for the communal effectiveness index toward plastic money acceptance.

In **communal service**, Active service programs help people to serve correctly. Their major components are low interest and zero processing fee. These two factors are influenced communal service among all beneficiaries.

From Table 6.10, it is observed that the acceptance rate among the community of the Jangalmahal region is significantly high which is 86.1%. That means the rural people in India are appreciated the use of plastic money for their reutilized financial transactions.

6.6 Conclusion

In a nutshell, it can be concluded that demographic profile is an essential factor for the use of plastic money in the Jangalmahal region. Mostly, young and male people are using plastic money rather than cash. It is spotlighted

Table 6.10 Analysis of the communal effectiveness index.

Major factors	components	Agree	Average	ECS	CEI	% of CEI
Communal support (CS1)	Timesaving	189	0.921951	0.887805	0.860705	86.07046
	Reduce risk	169	0.82439			
	Portability	185	0.902439			
	Easy purchase	185	0.902439			
Communal security (CS2)	High accessibility	189	0.921951	0.850407		
	Technology	169	0.82439			
	Prestige	165	0.804878			
Communal service (CS3)	Low interest	173	0.843902	0.843902		
	Zero processing	173	0.843902			

that demographic profile (age, sex, occupation, monthly income, and educational qualification) is the most crucial factor for useful analysis and understanding the factors related to the use of plastic money. The perception of the users of the plastic cards is persuaded by the three essential factors, namely influencing factor, security factor, and finally cost factor constituent by nine identified variables. It is quite surprising that people of Jangalmahal being residents of a backward region are using plastic money influenced by variables like time-savings, portability, and easy purchase.

It is also remarkable that a good number of the people in Jangalmahal region are using the plastic card as an alternative to cash because it is issued by the financial institution free of cost or zero processing fee. When the institution charged any amount as processing fee, the perception of the plastic card users might be changed, and they will prefer liquid cash for the transaction.

The effectiveness of using plastic money in the Jangalmahal region highlighted the high degree of success in accepting the plastic money. Jangalmahal region peoples are receiving the cashless society concept to adapt to the modern culture. Their perception in the use of plastic money is that it gives them some extra facilities and gives additional values in society.

References

Annamalai, S., & Muthu, I. R. (2008). Retail transaction: Future bright for plastic money. In *Facts of you* (pp. 22–28). Kerala, India.

Bag, S., Aich, P., & Islam, A. (2020). Behavioral intention of "digital natives" toward adapting the online education system in higher education. *Journal of Applied Research in Higher Education*. https://doi.org/10.1108/JARHE-08-2020-0278.

Bag, S., & Omrane, A. (2020). Corporate social responsibility and its overall effects on financial performance: Empirical evidence from Indian companies. *Journal of African Business*. https://doi.org/10.1080/15228916.2020.1826884.

Barker, K. J., D'Amato, J., & Sheridon, P. (2008). Credit card fraud: Awareness and prevention. *Journal of Financial Crime, 15*(4), 398–410.

Bartlett, M. S. (1950). Test of significance in factor analysis. *British Journal of Mathematical and Statistical Psychology, 3*(2), 77–85.

Bernthal, M. J., Crockett, D., & Rose, R. L. (2005). Credit cards as lifestyle facilitators. *Journal of Consumer Research, 32*(1), 130–145.

Bisht, A., Nair, P., Dubey, R., & Hajela, T. (2015). Analysis of use of plastic money: A boon or a bane. *SIMS Journal of Management Research, 1*, 5–12.

Brito, D. L., & Hartley, P. R. (1995). Consumer rationality and credit cards. *Journal of Political Economy, 103*(2), 400–433.

Chakravorti, S., & Emmons, W. R. (2001). *Who pays for credit cards? Federal Reserve Bank of Chicago*. Policy Studies EPS.

Chin, K. Y., Yi, W. C., Kwan, C. L., & Si, C. M. (2018). An empirical investigation on credit card repayment pattern of academicians in Malaysia. *International Journal of Banking and Finance, 13*(2), 21–42.

Dhanda, K., & Arora, U. (2017). Genesis of cashless society: A study on growing acceptability of plastic money. *International Journal of Advanced Research and Development, 2*(5), 478–485.

Feinberg, R. A. (1986). Credit cards as spending facilitating stimuli: A conditioning interpretation. *Journal of Consumer Research, 13*(3), 348–356.

Hafalir, E. I., & Loewenstein, G. (2012). The impact of credit cards on spending. *Bulletin of Business Research, 50*(2), 5–8.

Hausman, A. (2000). A multi method investigation of consumer motivations in impulse buying behavior. *Journal of Consumer Marketing, 17*(5), 403–426.

Kaseke, N. (2012). Cash or plastic money—An investigation into the payment mode post multicurrency period in Zimbabwe. *International Journal of Advanced Research in Management and Social Sciences, 1*(6), 102–119.

Khare, A., Khare, A., & Singh, S. (2012). Factors affecting credit card use in India. *Asia Pacific Journal of Marketing and Logistics, 24*(2), 236–256.

Lee, Y. H., & Huang, Y. L. (2011). Do you have credit cards? The expansion of the credit card market in Taiwan. *Applied Economics Letters, 18*(17), 1639–1644.

Malhotra, N. K., & Dash, S. (2010). *Marketing research: An applied orientation*. 6/e India: Pearson Education.

Manivannam, P. (2013). Plastic money cash less payment system. *International Global Research Analysis, 2*(1), 11–12.

Mester, L. (2006). *Changes in the use of electronic means of payment: 1994–2004* (pp. 26–30). Federal Reserve Bank of Philadelphia Business Review (Q2).

Ming-Yen Teoh, W., Chong, S. C., & Mid Yong, S. (2013). Exploring the factors influencing credit card spending behavior among Malaysians. *International Journal of Bank Marketing, 31*(6), 481–500.

Nayak, T. K., & Agarwal, M. (2008). Consumer's behaviour in selecting credit cards. *Journal of Services Marketing, 12*, 49–59.

Nga, J. K. H., Yong, L. H. L., & Sellappan, R. (2011). The influence of image consciousness, materialism and compulsive spending on credit card usage intentions among youth. *International Journal of Business Administration, 12*(3), 243–253.

Park, H. J., & Davis, B. L. (2005). Fashion orientation, credit card use, and compulsive buying. *Journal of Consumer Marketing, 22*(3), 135–141.

Patel, B., & Amin, U. (2012). Plastic money: Roadmay towards cash less society. *Paripex—Indian Journal of Research, 1*(11), 95–99.

Phau, I., & Woo, C. (2008). Understanding compulsive buying tendencies among young Australians: The roles of money attitude and credit card usage. *Marketing Intelligence & Planning, 26*(5), 441–458.

Rotchanakitumnuai, S., & Speece, M. (2004). Corporate customer perspectives on business value of Thai internet banking. *Journal of Electronic Commerce Research, 5*(4), 270–286.

Soman, D., & Cheema, A. (2002). The effects of credit on spending decisionss. The role the credit limit and & credibility. *The Journal of Marketing Science, 21*, 32–53.

Worthington, S., Stewart, D., & Lu, X. (2007). The adoption and usage of credit cards by urban-affluent consumers in China. *International Journal of Bank Marketing, 25*(4), 238–252.

CHAPTER 7

Paradigms of wisdom economy for manufacturing in a post-pandemic world

Arup Barman
Department of Business Administration, Assam University, Silchar, Assam, India

7.1 Introduction & history of manufacturing management

The simple concept of manufacturing is a group of activities undertaken by a person or a company that makes goods for sale. The person or the company who undertakes activities for manufacturing is the manufacturer who supplies the goods to the distribution center; they may be a wholesaler, retailer, or an exporter. From the meaning of manufacturing and manufacturer, we can define the term "manufacturing management." In simple parlance, planning, organizing, controlling, and directing activities of manufacturing of goods or produce are known as manufacturing management. Once the manufacturer decides to sell goods to the seller or the consumer, from that point, the responsibility of manufacturing management starts.

Manufacturing as a series of activities is carried out through processes. A process any activity or group of activities that takes one or more inputs transforms them to one or more outputs. The outputs could be for external customers or internal customers. External customers may use the product for sale to others or direct consumption. Internal customers may use it for the next process to give a shape of the final output. Manufacturing processes in other words are transformation processes that change the materials or inputs or raw materials.

Frederick Winslow Taylor in short F. W. Taylor was a pioneer who contributed to developing the concept of manufacturing management. F. W. Taylor contributed scientific management, scientific manufacturing process management, process improvement, and design. F. W. Taylor proposed that manufacturing managers need to take the responsibility for developing manufacturing methods. Frank Galbraith contributed by studying motion study in manual work or machine-operated work. Frank Galbraith with Lillian Galbraith conducted a fatigue study of machines operated and

Green Production Engineering and Management
https://doi.org/10.1016/B978-0-12-821238-7.00006-3

Table 7.1 Contributors to manufacturing management.

Theorists	Main contribution	Contribution(s) to production management
F. W. Taylor	Scientific management	Manufacturing process management, process improvement, and process design
Frank Galbraith	Scientific shop floor management	Motion study
Frank and Lillian Galbraith	Human impact on manufacturing operation	Fatigue study
Henry Ford	Production system	Assembly line in operation
Henry Gantt	Scheduling production activities	Gantt chart
Harry Emerson	Efficiency and industrial engineering	Efficiency of manufacturing system
F. W. Harris	Theory of batch processing	Batch quantities in production and purchase
Walter Stewart	Statistical thinking in control	Statistical process control
Hawthorne	Hawthorne experiment	Impact of psychological variables and behavioral variables

conjectured the negative consequences in manufacturing operation. They proposed ways to prevent fatigue and the negative consequences thereof (Table 7.1).

Manufacturing management got its momentum with the development and integration of operation research (OR). OR helped the manufacture to understand and optimize the system of manufacturing better. Further integration of OR led to advanced system implementation like manufacturing resource planning (MRP) which was extended to enterprise resource planning (ERP).

Till the end of the 20th century, manufacturing management not only locked up with US-based techniques and theories but also plastered by many Japanese techniques. The major Japanese concepts—total quality management (TQM), total productive maintenance (TPM), and total cost management (TCM)—became the strategies for production, manufacturing, and organizational management. Later, just-in-time (JIT) or lean systems became the best practice for production systems. New technologies dominated along with the advancement of production engineering. The theory of business process reengineering (BPR) helped in utilizing the full power and the potential of the adopted new technologies more. Along with the spread of the internet and the adoption of new technologies, the manufacturing process could acquire the ability to look like superior, and superior systems using operation research models and system dynamics models with the ability to access data anywhere using internet-based communication made coordinated and distributed global facilities. With the development of technology and strategies over the last two decades (during 1990–2010), there was a rapid succession of manufacturing management concepts. From the year 1984 to till 2012, the "theory of constraints (TOC)" was a dominant innovation fling for manufacturing management (TOC Institute, n.d.). During that period, the concepts like a focused factory, just-in-time manufacturing, concurrent engineering, total quality management, supply chain management, flexible manufacturing, lean production, mass customization, etc., are not last, but had pushed forward manufacturing management to a high-tech economic activity.

7.2 Manufacturing in knowledge economy

Manufacturing activities just after the brick-and-mortar economics began to hold close the knowledge economy paradigm. Production and manufacturing purely became very specialized activities with high-level automation and mechanization to fulfill the demand and supply. Consumers' demand was an unsatiate psychological attribute generated by information and knowledge. Due to the infusion of information, together with production, the attempt of a marketeer to differentiate the product with the help of knowledge manipulation near the consumer becomes insatiable. The producers used to take the advantage of individual demand—national, regional, and even international. Thus, knowledge (K) components have been hiddenly added with the traditional inputs, such as land (L), labor (La), capital ©, and organization (O). The production function of the knowledge-led economy was comprehended as a mix, i.e., $P=f(L, L+k, C+k, O+k)$. Deploying these manipulative inputs, combination production management became a multifaceted production gimmick as revolutions in the name of mass manufacturing, just-in-time (JIT) manufacturing, quality-led manufacturing for competition, and finally, technology-integrated economic activities.

7.3 Knowledge economy paradigms

As stated earlier, knowledge is a critical input mix with all traditional input combinations of production and manufacturing. A few distinctive characters of knowledge economy controlled every activity in any economy of the world. These characteristics are exerted as knowledge economy paradigms and they are:

(1) The *knowledge economy always wants more,* i.e., never-satiating wants and needs are the prime character. Knowledge economy values only accumulation.

(2) As knowledge cannot be assessed easily, the knowledge economy *demands qualifications.* Performing work in a knowledge economy needs skilled workers formally qualified to perform.

(3) The knowledge economy is built around ideas and intellectual capital. In common parlance, knowledge economy is *technology-driven* and the driving force of the technology-led economy is determined by the pace of development of ideas and intellectual capital.

(4) All sectors of knowledge economy are responsive to ideas, intellect, and level of technology; hence, the *knowledge economy is innovative* to adopt high technology and change.

(5) As the knowledge leverage is the competitive enabler in the economy, the knowledge economy is competitive. The economic value of knowledge is recognized in the knowledge economy.

(6) As the knowledge economy wishes to prove knowledge superiority, establishing superiority in a natural context is difficult. Due to complicacy to establish knowledge superiority, the component of power gets mixed through politics and society. Therefore, the *knowledge economy is political by nature and autocratic in spirit.*

The above paradigms dictate the scenario of manufacturing management for the entire world. To demonstrate behold knowledge as superior, along with the speed of globalization, manufacturing economies used heinous, health crisis and geopolitical strategies during the peak of the knowledge era.

7.4 Cessation of knowledge supremacy

Knowledge is useful information and facts. Once selected and analyzed, the information and facts, when processed, produces some results. By deploying the knowledge economy paradigm with manufacturing, some economies even dreamed to be a superpower in the world. Before COVID-19, the supremacy of knowledge economy had signaled a muffling picture. Dobson (2017) in headlines stated "stop talking about the knowledge economy, start building wisdom economy." Dobson stated "we know we're in the middle of seismic shifts in the way the world operates." He articulated a few speculations based on the development and trends spawned by the operation of the knowledge economy in the world. Dobson in his language expressed "we don't know where they'll end up, or where any of us will be when the dust settles if it ever does. Will we have a job? A pension? A home? Will there be someone to care for us in old age?" This is the pride of knowledge-based manufacturing economy because the knowledge workers in manufacturing use information to achieve tasks of manufacturing to fulfill the target. To make single-aimed decision to manufacture goods, knowledge workers rely on education and training those are very specialized or aimed to fulfill the target through knowledge. Knowledge workers in manufacturing work rely on the rules dictated throughout the education they gained from the training for manufacturing.

The knowledge economy never uses moral, ethical, intellectual, philosophical, and experiential insights blending with information for doing things. Either knowledge-led manufacturing is to rely on industry-wide observation or apply any principle to seamlessly navigate emerging situations (Jakovickas, 2019). It is implicit that insightful workers possess the ability to synthesize all rudiments of knowledge and experiences and wherever they work, they can "convert them with insights, that give them the more profound capacity to understand the connections, and in reality, the relationships between things with the meaning of life (Roberts, 2019)." Considering the inherent attributes of manufacturing philosophy in the knowledge economy distasted the humane economists. Therefore, it is realized that it is time to give a full stop furthering the knowledge economy in the world.

7.5 Covid-19 on manufacturing

COVID-19, a laboratory-produced virus, used by China as a weapon for conspiracy against the global economic war was spread from the Wuhan Lab of China during the last part of September 2019 to satisfy the heinous ambition to make the tag "Made in China" as the superpower in the globe. This was the ambition of China to posit its productions and manufacturing at the ultimate height of the global business. Chinese acclimatization to the knowledge economy and consequent arrogance impact the human dwelling as the earth for dead bodies and domes. This was a weaponless massive disruption to the social and communities, economic life, technological life, and political lives of humans in the world. COVID pandemic forced the manufacturing industries and the businesses for a radical reaction not to survive through but to ensure the survival of life by instantaneous changes. Factory closures, short-time work, and salary and job cuts are significantly impacting lives around the world. However, there is also room for opportunity, as manufacturing can use this time to focus on projects that were halted, clean up IT and supply chain inefficiencies, and reset their product catalog and strategy. COVID-19 enforces midterm actions or interventions as chaos in long term, availing the opportunities in industrial manufacturing amid all disruptions.

COVID-19 forced us to think that human beings should start thinking from ground zero which have been failed in many areas. The ground zero thinking areas are economics, mathematics, moral philosophy, and history (Hart, 2020). The outbreak of COVID disaster left no one in the world to feel that "everyone is vulnerable, everyone is facing a high degree of uncertainty, and the consequent circumstances in the world are very complex at all levels," there is no single solution to a problem, and all problems demand multiple answers and options due to ambiguity. In the IESE Insight, Business Knowledge (2020), it is stated that COVID-19 outbreak has strong VUCA situations—in the acronym referring to *volatility, uncertainty, complexity, and ambiguity*.

The search for solutions and applications for restoring the system entangling multiple issues with numerous doubts. All those disarrays twigged many promising suggestions to counter the chaos created by COVID-19. The twigging solutions for post-COVID-19 manufacturing are appealing epistemes, and they are agile manufacturing, flexible manufacturing, and digitalized and automated manufacturing (TULIP, 2020). Suddenly, the manufacturers had to forget the broader world, situations compelled to look inside the factory, and consequently, the whole manufacturing world had to militate against the rapid shift in production in its own way. Amid all meddling situations, society hoped "COVID-19 will end within months, and things will go back to normal. Many aspects of our society may never return to normal" (Kropenev, 2020). The global shutdown became history in the modern knowledge-based economy and in the history of manufacturing management by the impacts of demand and supply disruption and workforce availability simultaneously. To comprehend more, the effects of COVID-disrupted manufacturing can be categorized under two heads—short-term effects and long-term effects (Table 7.2).

Table 7.2 Effects of COVID-19 on manufacturing.

Short-term shifts	Long-term shifts
Demand increased and supply decreased due to • Panic buying of delivering vital goods like personal care, paper, pharmaceuticals • The dramatic drop in demand for other than above	The revival of automated domestic manufacturing Emphasis on robotics and digital application for productivity enhancement Decoupling supply chains with the existing technology, with geographical locations, and new economic collaborations
Disrupted supply chain—became volatile	Digitalization as a means for competitive advantage and use of artificial intelligence (AI) and internet of things (IoT) technologies
Social distancing and safety measures, the deficit of on-site labor	Scheduling on-site worker management by virtual and physical shifts
Remote work (for knowledge system)	Rapid adoption of remote diagnostic management and application of collaboration
Heavy demand for networking tools and infrastructures	Responding to the fundamental shifts of government policies and consumer behavior; response to geopolitical behavior and new normal local political behavior

7.6 Developing resilient system

The pandemic situation dictated to create a safe workplace, flexible technology, and rapid adoption, responsive to the demand system of infrastructure. Considering the requirement of social distancing in crowded operating workspace stipulating the adoption, application, and absorption of robots in many complex manufacturing processes, technology and infrastructure became more foreseeable. The episode of post-pandemic preparation in Mahindra & Mahindra witnessed that "rise of robots in automobile manufacturing was inevitable," and also stated as they "have automated the body shop, most of the paint shops and parts of the final assembly line and touchpoints need to get automated" (Bhattacharya & Phillip, 2020). Besides, industrialists are imagining to apply virtual reality (VR), augmented reality (AR), and mixed reality (MR); layering over reality to produce goods and entertainment will not be far from reality to cope up with the manufacturing process disruption (Robert, 2019). Interacting the complex event under business reality and the reliance of collective intelligence and cloud manufacturing become center-stage imagination. Knowledge-based manufacturing due to the impact of COVID-19 may further remain as the victim of technology. As manufacturing is understood as a sociotechnical system, the rigidity of the manufacturing needs to be busted down by integrating technology for intelligence. But, it is also understood that the combination of technology and intelligence together does not have the power of maintaining manufacturing resilience. We experienced how COVID-19 dictates to normalize the whole paradigms of manufacturing. Manufacturing management understands the urgency of being conscious of pragmatism and concern for the social world.

7.7 Wisdom economy

Manufacturing as the victim of emerging technology contradicts the universal wisdom and is a sociotechnical domain. Furtherance of discussion on manufacturing management under the ambit of technology integration, adoption, and absorption may lead to a qualm of deficiency that manufacturing under knowledge economy would not serve the human beings by the foundational meaning. It might be because pre-COVID economy was not that mature to explore on wisdom, but there is evidence of theoretical exploration on wisdom manufacturing. Xifan et al. (2014) investigated and conceptualized wisdom-based manufacturing as the synthesis of organizational wisdom, collaborative learning, and innovation and creativity. The implicit and explicitness to the holistic integration of organizational wisdom, collaborative learning, innovation/creativity, computers, and humans integrated into a system engineering may trigger a vast paradigm shift in the manufacturing world. Let us go to the basics of wisdom economy to understand the potential vigor in paradigm swing connecting to manufacturing in the future.

The nucleus of the wisdom economy designates "wise application of knowledge" as the key driver and "complex adaptive system" as the core metaphor (Findley, 2011). Economists continue to assert the creative power of human collaboration in generating new ideas, innovations, and discoveries that on the whole have improved our quality life and created enormous opportunities (Sunde, 2014). The quality of human life relies on wise production. The wisdom began to act only at a specific point. As a human when anyone realizes and starts to fear God, from that point onward wisdom starts working. A human person participates in the broader human family by his/her nature (Sunde, 2014) by the level of acquisition of wisdom. We understand that every individual family in an ideal sense is a rich expression of love and solidarity. Let us understand the consumers and the producers are two groups of members in a family bound by the market or by the society and both of them must witness the love and solidarity in the world. The juxtaposed points between love and solidarity work as corollary wisdom that may control demand, supply, production, and consumption, in other words, will be characterized as wisdom economy. What are those foundational attributes of wisdom economy?

7.8 Characteristics of wisdom economy

"What makes sense?" of the wisdom economy is a grave issue to ponder over. Deepening down in the philosophy of wisdom connotes that the phenomenon of the impermanence of anything in the world is a permanent phenomenon. As per the Buddhist dictum, "impermanence" says that everything on the earth changes; ignorance is the main cause of suffering. A human being cannot alternate many things in the earth so human beings at the time of an uncontrollable situation should adopt let-it-go or leave-everything-to-god feelings. A present moment spent meaningfully will beget a fine future moment (Sarma, 2020). So, humans follow the instructions of ultimate wisdom to live in the present. Wisdom economy will propagate material production by the constructs of abundance and will be controlled by the theory of constraints. If human beings (producers and consumers) remain ignorant, then they must suffer. To live in the present in the earth through demand and supply, the wisdom economy dictates to use ultimate wisdom suffice to present and future. The fundamental characters of the wisdom economy are explained in Table 7.3.

As there are issues to live in present through impermanence, the wisdom age proliferates complex adaptive system which makes a sense in the economy of wisdom age. The metaphors of the wisdom age would help us in understanding wisdom age manufacturing functions.

Table 7.3 Key attributes of wisdom economy (WE).

SL No	Key attribute (s)	Interpretations and implications
1	Concept of "enough or abundance"	**Wisdom** knows that a person's life does not consist of abundance of possessions. **Wisdom** understands prosperity as a state of sufficiency
2	Insistence for quality	Wisdom economy is reflective. The quality of anything counted is understood deeper, thoughtful, philosophical, and insightful as opposed to superficial
3	Collaborative	Wisdom economy is a collaboration and does not encourage much competition. They better ensure coexistence through collaboration. Producers and consumers collaborate to coexist
4	Local product	Wisdom economy emphasizes on crafting or creating a local product and known consumptions
5	Personal economy	Wisdom economy ensures higher personal welfare
6	Crowd view	Wisdom as a view from crowd point, "crowd wisdom" assumes that grouped people are collectively smarter than a single or individual expert

7.9 Metaphors of WE on manufacturing

Wisdom economy (WE) is metaphorical, meaning that it is emblematic and symbolic. The wisdom economy activities, e.g., manufacturing management are symbolized on every subactivity. The component "system" in economic activity obviously will be symbolized as a complex adaptive system. The way the complex adaptiveness is reflected in the ecology, market, or brain, the same way any system adapted for manufacturing must be reflexive of complexity and adaptiveness. Thus, many components of wisdom economy are metaphoric and can be understood with the following Table 7.4.

Wisdom as philosophical is critical. Similarly, capturing the traits of manufacturing activities with a defined set of metaphors is difficult to reflect wisdom-based manufacturing due to the overlying aspect of depth embedded inside the wisdom economy. Hence, further amplification on paradigms of wisdom economy's impact on manufacturing management provokes discussions on total value management, ecology environment and sustainability, chaordic and agility aspects, etc., which are needed to be considered even above the limits of indispensability.

Table 7.4 Metaphors of wisdom economy.

Components of WE	Metaphors of wisdom economy
System	Complex adaptive, e.g., ecology, market, or brain
Technologies	Nano technology, biomimicry, advanced energy
Tools for expanding consciousness	Social media, complex learning environment, complex and ever-changing knowledge will be an enabler to remain connected in the world
Productivity gains	Automates scientific, judicial, leadership or wise expert work
Change	Rapid change, transformational from unconscious to conscious state
Nature of knowledge	Wise application of knowledge
Role orchestrator	Interactor, challenger—designer or creator, inspirer activist, researchers, discoverer, production user
Production and manufacturing methods	Agile, adaptive, value add—high value add system, rapid prototyping, customizable by customer, anticipate own need and customize
Strategic focus	Whole system, multiple generations, paradigms and cultures
Shift of focus	From boundaries to horizons, i.e., prospect, possibility, perspective of product use
Principles of coordination	Facilitate what emerges, improvisation at local space, simulation, aims at maximum efficiency

7.10 Total value management through manufacturing

It is a commonsense matter that manufacturing management must correlate the consumption of manufactured goods to a better quality of human life referred to earlier. Furthermore, it is more than an ordinary sense that beyond the level of consumption may well have negative consequences (TVM, 2020) and can be substituted with "too much everything is bad." That means overconsumption cannot be a solution by the use of refined and costumed terminology, like "adaption." The basic wisdom that needs to be controlled by the guiding of "what impact was" will be controlled by "design what if," and the cause of action is "decide what to do?" Knowledge dictum of value management will get the next shift to total value management (TVM). The total value of production and manufacturing having cane weighed with 5Ps. So, the 5Ps as a mix in the manufacturing value management work in the following table.

5Ps for Manufacturing			
5Ps (Manufacturing Mix)	Value for Whom/What/Why?		Total/True Value
P1-Process	Engineering		Value for Human's Life
P2-Profit	Economic Goals	→	
P3-People	Families, Society		
P4-Place	Where People Live		
P5-Products	Quality of Life		

As value management is an umbrella term, it contains a lot of value engineering terminologies that reflect how much it is ethical and wise. Say, to create a value in manufacturing scarcity, economics and knowledge economy use the cost of reduction techniques. Cost reduction as a technique of value management may impact either positively or negatively. Value engineers must use the wisdom for value optimization by considering the 5 Ps of manufacturing. Going further on value management, more other inputs and assumptions may add the paradigm of manufacturing. They are:

(a) Value management is a creative solving process needing a methodical problem-solving approach. So, total value management demands creativity, cooperation, collaboration, and competition-free exchange for the solution of any problem relating to products and processes.

(b) Total value management is all about design. Designers in manufacturing connect the value strings of the product of services where to pin up. Collaborative designs may delve over critical aspects of the value intended to create through products.

(c) The aspects of value are more about value governance. In the process of governing the value, for governance, 5 Ps—true value matrix—may serve as underpinning principles and provisions.

(d) Value management in manufacturing is a transdisciplinary subject. So, mere professional excellence for manufacturing may not serve any good result. Professionalism expertise suffices in the knowledge economy, but in the wisdom economy more than professionals will be a major concern.

7.11 Technology and sustainability paradigm

The issue of sustainability in handling manufacturing operations during the last two decades had been giving a serious message of squirting of human collective wisdom. We would be able to experience the prolonged taste of reaching the seventh heaven only when we collectively follow those controlled maneuvering grounds of collective ecstasy through sustainability. The controlled maneuvering exercises would cover the controlled use of energy, resources, and environment used in the manufacturing of goods and services. Technocrats are suggesting hybrid equipment in electrifying, decarbonized fuel utilization, and use of digital technology for sustainable manufacturing operations. The industries of the world have to adopt hybrid technology, process, and systems in the future to come. The hybrid—the future already has begun (Somers, Spellman, & Witteveen, 2020).

The "hybrid" will dominate as well as revolutionize the whole manufacturing in wisdom-based econosphere. Hybrid manufacturing will leverage the most exclusive capabilities of both additive and subtractive methods having high-value precision in a single machine-based manufacturing system. It is assumed that hybrid manufacturing will not only reduce the manufacturing costs but also will increase the economic competitiveness of technology-based manufacturing. Hybrid manufacturing will have the capacity to integrate traditional manufacturing with high-profile machining (Strong, Kay, et al., 2018). Future manufacturers will be able to produce custom material and rapid prototypes vary with the hybrid operation. The hybrid manufacturing principle may ramp up in providing an opportunity for the new product development and delivery with better efficiency (DMS, 2018).

Moreover, the hybrid process of manufacturing by using the additive method holds the potential of reduction of wastage and thus would decouple the environmental loss and the value creation. Hybridization and additive technology would substantiate many potential sustainable benefits and can be sharpened out under three strands:

- **Improved resource efficiency** (Ford & Despeisse, 2016) with the help of redesigning the use phase(s) and the processes of manufacturing.
- **Extended product life** can be accomplished through technical approaches such as repair, remanufacture, and refurbishment and more sustainable socioeconomic patterns, such as person-product affinity, maintaining a close relationship between producers and consumers (Kohtala, 2015).
- **Reconfiguration of value chain** by shortening and simplifying, adopting more localized production, innovative distribution, and new collaboration (Ford & Despeisse, 2016).

7.12 Climate change paradigm

Manufacturing and greenhouse gas footprint is directly connected. Manufacturing in the future will be impacted by the potential climate change regulatory regime. Economic activities in the future will be impacted by the nonfamiliar chaos created by climate change. Rifkin through his review foresees the nonfamiliar consequences of climate change. Rifkin predicted many worldwide happenings due to climate change, and they are like shrinking food availability, superstorms, high-intensity water-related events, extreme wind, changes in climate patterns, droughts, uncontrollable wildfires, loss of biodiversity, etc. (Rifkin, 2017). Days have come to close down the present fossil fuel economy in the world. The value of renewable energy has the potential to enter to energy, product, market, and value chain. At this moment, mere thinking about carbon reduction and manufacturing will not be that much mature to prognose human and economic activities that would rebalance the climate-related happiness in the world. It would be extremely foolish if the human being does not understand the signal of climate change, perhaps the human race is going to create an imminent doomsday.

7.13 VUCA paradigm

The world under the tumultuous situations due to the present COVID impact could help us imagine the 4 situations, and they are volatility, uncertainty, complexity, and ambiguity, in short VUCA. Business world in 2020 could have realized the volatility in the context of society, economics, and geopolitics which are still jerking through instability and through unexpected changes (Barman & Potsangbam, 2017). Disruption by COVID-19 created a lot of uncertainties. Scientists, insdustrialists, businessmen, and politicians could not visualize clearly due to the foggy situation created by the unseen virus. Everyone had been suffering an opaque situation due to the lack of understanding of what will cause a change for next normality and how it will be? Economic phenomenon occurs under open, complex, and dynamic interplay of various factors. As manufacturing management needs to undertake evermore in an open environment, due to the complexity of interrelation, the whole system and processes need to operate under the undefinable complexity. When the world remains under the situation of unknown unknowns or in other words when there are no obvious precedents, then we call it an ambiguous (synonyms are indistinct, hazy, unclear) situation. COVID pandemic created a lot of ambiguities in manufacturing and production due to the competing demand solution in the short run as well as for the long run. VUCA situations experienced by the world due to COVID-19 "demand a strategic shift that needs to be grounded by scientific, realistic, practical, and humanistic" (Barman & Potsangbam, 2017).

As the post-COVID industry and manufacturing will face continuously VUCA situations, a resilient or agile system of manufacturing would be considered as a direct solution. A knowledge-based view ofVUCA response is to develop the absorptive capacity of the manufacturing system. The knowledge-based view of absorptive capacity highlights the role of absorptive capacity in "developing knowledge, promoting organizational learning, enhancing open innovation, managing alliances, creating strategic variety, and impacting financial performance" (Lichtenthaler, 2016). In promoting learning-based manufacturing, the perfect fit for the VUCA-responsive system will rest on design thinking. Design thinking and design attitude in connection with the complex external environment would support in "operating in VUCA environments focusing organizations on cocreative and proactive learning by building design thinking structures to preempt the VUCA environment" (Cousin, 2018). Learning to produce things proactively for the VUCA situations ensuring that human well-being would glorify the wisdom economy in the future. Putting an insight gained from VUCA parlance, the wisdom economy is going to ensure an opportunity in and out of chaos.

7.14 Holistic econosphere paradigm

Let us understand the meaning "this is a world with the most fundamental aspect of everything" (Thomas, 2009) of holistic paradigm. Under this frame of reference, a manufacturer needs to make their intelligent choice from what we are—a producer for human well-being or something else? Similarly, every human being can decide a course of action(s) with an intelligent choice under the holistic frame of reference. Thomas (2009) has forwarded a wise dictum "we are living, breathing, working with a self-regulating holistic econosphere, whether we know it or not." Only we need to regulate ourselves while in manufacturing under the frame of truism of holistic econosphere.

We are expecting that the entire human race to collaborate directly with one another will democratize our own economic life. Future technological realm of predicting digital integration would help us connect producers and consumers, together "prosumers" (Rifkin, 2017). The evolving internet of things (IoT) will enable conventional produce, businesses, as well as millions of prosumers to decide and democratize about their manufacturing under the holistic econosphere. The zero marginal cost economy will grow more and more, boost up the sharing economy, and influence the trend of decision under holistic econosphere. The information technology-based sharing economy may raise further doubt—"where is the wisdom when we lose ourselves on information and knowledge? Unless we operate anything keeping in mind for the world, the information economy may create a dark and disconnected world.

7.15 The innovation perplexity

Manufacturing activities cannot be operated efficiently without compatible standardization or innovation. In contrast, too much standardization till the point of compulsive stoppage of innovation may lead to stultification and a push to losing out in the nimble competition. Excessive innovation without care or standardization may result in quality control further leading to manufacturing chaos. Managing manufacturing normality requires coupled innovation and standardization drive; the logic behind this is if we go all-in for innovation to exclusion of standardization, we may create many big problems. The challenges of manufacturing management will be "to explore how we could draw the strengths of each of the interdependent poles while simultaneously avoiding the problems that each brings when it is overemphasized" (Shenchez, 2020). This would bring back to face the question relating to the pandemic at a global, national, and local scale.

Let us remember "wisdom economy is innovative and reflexive." Reflexion does not drive out any innovation. Wisdom economy promotes responsive innovation in any sector. Manufacturing activities must not be under the perplexity on innovation but integration of reflexivity on innovation.

7.16 Linking paradigms and conclusion

The logic and philosophy behind wisdom economy principles with manufacturing management may look subjective. We must go ahead with the objectivities which were embedded in the knowledge economy in the world. It is time to make our economy shift along with the shift of industries. Before deepening down minds in analysis and the effects of impending full-fledged wisdom economy on the manufacturing management, "we need to value wisdom as much as we have come to value knowledge and make it central in our decision-making and leadership." Here, the crux of value statements of the wisdom economic council's need is to be taken into consideration.

1. We must believe in the global healthy business climate for the prosperous global economy, which is possible to make sustainability through the healthy and vibrant quality of life for all human beings.
2. We must demonstrate the highest level of ethical business practice and should act for peace and humanitarian causes locally and globally.
3. Every action and strategy meant for the economy must adopt a more inclusive approach embedded with wisdom for sustainability, insight, context, and commonsense as the clichés of the wisdom-based economy.
4. We must promote the consciousness for the needs of others before devising out the solution or displaying intellectual brilliance.

The days have come that the human being will operate everything under the ambit of singularity—the wisdom-connected globe. And the global wisdom may put everything in the cloud as well as everything will be among the crowd. Wisdom economy in the future will accelerate the wisdom of crowd which will set in motion to the next stage of economic development. Anyone whether the producer of industrialist can participate, contribute, "manage," lead, or govern "anything," anywhere, anytime, involving anyone (Cake, 2020). The only requirement will be "wisdom".

Under the wisdom economy, manufacturing management will be impacted by total value and true value management, technology, sustainability, climate change, VUCA, and innovation perplexity. Paradigms above the whole aspects will be controlled wisdom for the earth and human being and would be a deep-seated determinant of all activities of human beings on the earth. The post-pandemic manufacturing world may strike the balance by considering the short-term and long-term perspective of human welfare.

References

Barman, A., & Potsangbam, C. (2017). Shifts of Strategic Paradigms in the VUCA World—Does "outside the box thinking" a meaningful cliché for the business world? In *Managing Change and Creativity and Innovation in a Dynamic Environment at: Sonarpur, Kolkata.* Volume: 7th and 8th April 2017 https://www.researchgate.net/publication/315912311.

Bhattacharya, R., & Phillip, L. (2020). *Automation key to post-pandemic production; the economic times, APR 27, 2020.* https://economictimes.indiatimes.com/news/company/corporate-trends/automation-key-to-post-pandemic-production/printarticle/75395665.cms. 1 (Browsed on May 5, 2020).

Cake, M. (2020). *Organisation wisdom: digital freedom challenge. https://www.mixprize.org/hack/organisation-wisdom?_ga=2* (Browsed on June 3, 2020).

Cousin, B. (2018). *Design Thinking: Organizational Learning in VUCA Environments.* Research Article: 2018 Vol: 17 Issue: 2 https://www.abacademies.org/articles/design-thinking-organizational-learning-in-vuca-environments-7117.html. Browsed on June 4, 2020.

Dobson, J. (2017). *Stop talking about the knowledge economy. Start building a wisdom economy. Smart cities dive.* https://www.smartcitiesdive.com/ex/sustainablecitiescollective/stop-talking-about-knowledge-economy-start-building-wisdom-economy/9225/. ¼ (Browsed on May 18, 2020).

Findley, J. (2011). *Opportunities in the chaos, in the Blog Post: The Wisdom Economy.* http://thewisdomeconomy.blogspot.com/ (Browsed on May 21, 2020).

Ford, S., & Despeisse, M. (2016). Additive manufacturing and sustainability: An exploratory study of the advantages and challenges. *Journal of Cleaner Production, 137*(2016), 1573–1587.

Hart, D. (2020). *The urgent need for intellectual change.* aaier.org/article/the-urgent-need-for-intellectual-change (Browsed on 30th April, 2020).

Jakovickas, C. (2019). *Evolve from knowledge worker to wisdom worker.* (in https://thebusinessjournal.com/blog-evolve fro-knowledge-worker-to-wisdom-worker) and in Url Https://Thebusinessjournal.Com/Byline/Cassidy-Jakovickas/ Browsed on May, 20,202.

Kohtala, C. (2015). Addressing sustainability in research on distributed production: An integrated literature review. *Journal of Clean Production, 106*, 654–668.

Kropenev. A, (2020), What will manufacturing's new normal be after COVID-19? History teaches; Industry Week-2020; https://www.industryweek.com/technology-and-iiot/article/21129334/what-will-manufacturings-new-normal-be-after-covid19.

Lichtenthaler, U. (2016). Absorptive capacity and firm performance: An integrative framework of benefits and downsides. *Technology Analysis & Strategic Management*, 664–676.

Rifkin, J. (2017). *How the third industrial revolution will create a green economy; HUFFPOST.* Updated Dec 06, 2017 https://www.huffpost.com/entry/third-industrial-revolution-green-economy_b_8286142 (Browsed on June 3, 2020).

Robert, R. (2019). *VR, AR, & MR in education.* https://medium.com/awecademy/vr-ar-mr-in-education-e789bfe5b869; (Browsed on May 17, 2020).

Roberts, R. (2019). *Manufacturing wisdom: big data in education.* https://medium.com/awecademy/manufacturing-wisdom-big-data-in-education-42b8a3c7fdde (Browsed on May 18, 2020).

Sarma, R. (2020). *Buddhist wisdom in the outbreak of Covid-19; North East Now (NNN).* https://nenow.in/opinion/buddhist-wisdom-in-the-outbreak-of-covid-19.html (Browsed on, May 20, 2020).

Shenchez, A. (2020). When There Is No One Answer to Pandemic Problems. *CEOWORLD Magazine.* https://ceoworld.biz/2020/05/18/when-there-is-no-one-answer-to-pandemic-problems/ (Browsed on May, 18, 2020).

Somers, K., Spellman, E., & Witteveen, M. (2020). Hybrid equipment: The first step to industry electrification; re-imagining industrial operations. *McKinsey Quarterly*, (May 2020).

Strong, D., Kay, M., et al. (2018). Hybrid manufacturing—Integrating traditional manufacturers with additive manufacturing (AM) supply chain. *Additive Manufacturing, 21*, 159–173.

Sunde, J. (2014). *The economy of wisdom: How knowledge empowers service and stewardship.* Action Institute Powerblog. https://blog.acton.org/archives/74664-economy-wisdom-knowledge-empowers-service-stewardship.html (Browsed on May 18, 2020).

Švarc, J., & Dabić, M. (2017). Evolution of the knowledge economy: A historical perspective with an application to the case of Europe. *Journal of Knowledge Economy, 8*, 159–176.

Thomas, C. (2009). *The econosphere: What makes the economy really work, how to protect it, and maximize your opportunity for financial prosperity* (p. 2). FT Press.
TOC Institute (Theory of Constraint Institute), n.d. https://www.tocinstitute.org/history-of-toc.html (Browsed).
Xifan, Y., et al. (2014). Towards a pearl of wisdom manufacturing vision. *International Journal of Computer Integrated Manufacturing, 2014*. https://doi.org/10.1080/0951192X.2014.972462.

Web sources

DMS. (2018). *The rise in hybrid manufacturing*. https://dmscncrouters.com/the-rise-in-hybrid-manufacturing/ (Browsed on June 3. 2020).
IESE Insight, Business Knowledge. (2020). *Keys to beat volatility, uncertainty, complexity and ambiguity*. https://www.ieseinsight.com/docImpresion.aspx?id=2295&idioma=2 (Browsed on 20/05/2020).
TULIP. (2020). *Manufacturing before, during, and after COVID-19 (April 7, 2020 in COVID-19 response)*. https://tulip.co/blog/category/covid-19-response/ (Browsed on May 20th 2020).
TVM, True Value Matrix. (2020). http://www.truevaluemetrics.org/DBadmin/DBtxt003.php?vv1=L0700-TVM-PRODUCTS-01a (Browsed on May 25, 2020).

Further reading

Narayan, R. (2020). *Management Theory Review (Blog)*. http://nraomtr.blogspot.com/search/label/Digital%20Transformation (Browsed on May 21, 2020).
WISDOM ECONOMY. http://www.wisdom-brains.com/wisdom-economic-council/.

CHAPTER 8

Desertec initiative in Algeria: Challenges and barriers

Mohammed El Amine Abdelli[a], Aied Malika[b], Arabeche Zina[c], Soudani Ahlem[d], and Nedjari Fatima[e]
[a]University of Western Brittany-UBO, Brest, France
[b]University of Lounici Ali de Blida 2, El Affroun, Algeria
[c]Center University Ahmed Zabana Relizane, Relizane, Algeria
[d]University of Guelma, Guelma, Algeria
[e]Djillali Liabes University, Sidi Bel-Abbes, Algeria

8.1 Introduction

In worldwide, there is a high growth in the share of renewable energy (Wüstenhagen, Wolsink, & Bürer, 2007). For many years, the world has witnessed environmental pollution caused by the emission of harmful gases from chemical waste to factories. After the crises that accompanied the prices of fossil energy and the depletion that comes to it, increased interest in renewable energy sources has become a new topic for the economies of the energy world, allocating considerable funds to invest in this area, and among the countries that made very great efforts to develop these energies, European countries are in the lead based on their fruitful experiences and figures achieved in the exploitation of renewable energies of all kinds—solar, wind, hydropower, geothermal energy, and bioenergy—and through this study, we learn about the possibilities of European and African countries in numbers and analysis. From the above, we formulated the following question: what is the importance of investing in renewable energies for Europe and African countries? So, we divided the research into two main axes (02): the conceptual framework for renewable energies and an analytical study of renewable energy's reality in Europe and Africa. Is the application of the Desertec project in Algeria realistic or just expectations?

Green Production Engineering and Management
https://doi.org/10.1016/B978-0-12-821238-7.00009-9

8.2 The conceptual framework for renewable energies

8.2.1 Renewable energy concept

Renewable energy is a general concept used to describe energy output (Harjanne & Korhonen, 2019) while Lowitzsch, Hoicka, and Van Tulder (2020) define renewable energy clusters are characterized by the complementarity between different energy sources, stability, interconnectivity between other actors, and bidirectionality of energy flows. The International Energy Agency's concept is "renewable energy is formed from the energy sources produced by the spontaneous paths of nature, such as sunlight and wind, which are regenerated in nature at a higher rate than their consumption." Kaya (2006) defines the energies as renewables (solar, hydroelectric, biomass, wind, ocean, and geothermal). The IPCC conceptis "renewable energy is all energy that is sourced from geophysical or biological solar, which is regenerated in nature at an equivalent rate or greater than its usage rate and is generated by successive and continuous currents in nature at a more equal or more equal rate of use than successive trends and generated by successive trends. Such as heat and electric power and kinetic energy using multiple technologies that allow the provision of energy services from fuel and electricity." Concept of the United Nations Environmental Protection Program is "Renewable energy is an energy that does not come from a fixed and limited reservoir in nature that is periodically regenerated faster than the frequency of consumption and appears in the following five forms: biomass, sunlight, wind, hydropower, and underground energy."

8.2.2 Advantages of renewable energy use

There is a range of advantages that can be included as follows:

Energy consumption is the growth (Pacesila, Burcea, & Colesca, 2016). Renewable energy is inexhaustible, giving clean, waste-free energy. It aims first to protect human health and preserve the natural environment; it has a low production cost, improves human lives, reduces poverty, and creates new jobs. Reduced number and severity of natural disasters caused by global warming and the nonformation of acid rain damage all crops and life forms. Reducing the formation and accumulation of harmful waste in all shapes and protecting all living organisms, especially endangered species, protect groundwater, seas, and fisheries from pollution. Contributing to food security and increasing the productivity of crops as a result of their elimination from pollutants, it can be said that there is a trend in various developed and developing countries to develop policies to utilize and invest in all kinds of renewable energy as a way to maintain human health on the one hand and preserve the environment on the other, in addition to finding other sources and forms of energy that have the potential to continue, regenerate, and save at lower costs.

8.2.3 Renewable energy sources

8.2.3.1 Solar energy

The sun is a significant source of abundant, inexhaustible energy (Kabir, Kumar, Kumar, Adelodun, & Kim, 2018). It is the emitted light and heat from the sun that man has harnessed for his benefit since ancient times using a set of technology methods that are continually evolving. Most of the renewable energy sources available on the surface of the Earth are attributed to solar radiation in addition to secondary energy sources such as wind, wave energy, hydropower, and biomass; it is crucial here to mention that only a small part of the solar energy available in our lives has been used in our lives.

8.2.3.2 Wind power

The usage of wind power and other forms of renewable energy is developed at a remarkable pace (Nelson & Starcher, 2018). The generated energy is the movement of giant propellers installed on poles in high places by air. In addition, the wind produces electric power by the propellers. When the wind passes over the fan blades, it generates its cycle, and this rotation turns on the turbines, producing electrical energy.

8.2.3.3 Hydropower

Hydropower was the world's leading clean energy source power, accounting for as much as 71% of this product as of 2016 (Moran, Lopez, Moore, Müller, & Hyndman, 2018). It is the energy derived from the power of water, produced in various ways, including the energy generated from the flow of water or its fall in the case of waterfalls, drop water is the case in dams, the power of waves in the seas where the waves arise as a result of the movement of wind and do it on the waters of the seas, oceans, and lakes, and from the direction of waves create stoppable energy and convert into electrical power, where the waves are produced usually, the estimated capacity of 10–100 kW/m of the beach, the energy generated by the movements of the tides and islands in the water, as well as the energy generated by the thermal differences of the water layers of the temperature difference between the upper and lower layers of water that can reach a difference of 10°C.

8.2.3.4 Underground geothermal energy

It is a clean and renewable alternative source of high thermal energy of natural origin stored in the ground, where it is estimated that more than 99% of the earth's mass is rocked above 1000°C and the temperature increases by increasing our depth in the earth's underground at a rate of about 7°C and 2°C per 100 m in depth, i.e., it reaches an average of 27°C at a depth of 1 km or 55 at a depth of 2 km and so on. This thermal energy is mainly used in electricity generation, and this requires drilling of deep pipes that can reach about 5 km and sometimes use hot water for heating, and sometimes found at a depth of 150 m or sometimes in certain areas in the form of hot springs extending to the surface of the earth.

8.2.3.5 Biomass energy

Today, biomass is the world's most used source of sustainable energy (Toklu, 2017). Fast-growing trees, grains or vegetable oils, or agricultural residues may be energy extracted from organic matter from seeds, animal compost, waste, agricultural waste, or seeds used in the processing of biomass energy. Goods such as methane, alcohol, and steam are produced by various methods of processing biofuels such as direct or indirect combustion, fermentation, or distillation methods… and any of the prior techniques. And chemical fertilizers and ethanol gas are some of the better fuels produced in special pits to raise methane from the animal and agricultural mass, which is a flammable gas.

8.3 The reality of renewable energy use in Europe and Africa

8.3.1 The total renewable energy in euro-African countries (2010–19)

Table 8.1 shows the production of renewable energy during the period from 2010 to 2019 for African, European, and other countries of the world, and what draws the attention is the African countries despite the enormous environmental potential that they have in Sahara, which extends over a vast area and is the hottest desert in the world and has the hottest six regions of the world with the largest underground water reserves in the world, as well as the largest reservoir of fossil energy in the world, in addition to regional diversity and water bodies from the Mediterranean sea to the Red Sea to the Atlantic Ocean, where it has the longest river in the world, the Nile River, which originated in Lake Victoria and is the second tropical lake of freshwater in the world in terms of an area of 68,870 km^2 in Tanzania, but energy

Table 8.1 The total renewable energy in Euro-African countries (2010–19).

CAP (MW)	2010	2011	2012	2013	2014	2015	2016	2017	2018	2019
World	1,226,853	1,332,236	1,445,233	1,563,196	1,692,680	1,846,060	2,008,547	2,182,449	2,361,061	2,536,853
Africa	27,329	27,775	28,748	30,944	32,965	35,288	37,950	42,724	46,462	48,443
Europe	322,167	360,937	395,281	420,030	440,846	465,097	488,661	513,087	537,993	573,266

production renewable swells compared to the available resources and is still a fertile field for exploitation, with the production reaching 27,329 MW in 2010, while European countries had double output in the same year, reaching 322,167 MW, and the rest of the world by 2013, solar power production in African countries had some height of 30,944 MW, and in contrast to European countries, their show was 420,030 MW, also increasing global output by 1,563,196 MW, and by the beginning of 2017 African countries produced 42,724 MW while the European countries accounted for 513,087 MW, bringing the world's production to 2,182,449 MW. Globally, it was 2536,853 MW. Through the above, we conclude the following:

- The desire of governments has played a significant role in exploiting renewable energy.
- Countries' political stability would have saved vast amounts of armaments for the benefit of industry and renewable energy.
- Research and development interest would have provided the environmental technologies needed to exploit renewable energy.
- Attention to human competencies would have allowed the emergence of innovative and valuable ideas that show their value in the patents designed for this matter.
- The provision of laboratories and research institutes specializing in renewable energies would have allowed for innovation characterized by continuous learning and thus the optimization of available resources.

Fig. 8.1 shows the production of renewable energies in both African, European, and other countries globally, where global production and European production appear to be on the rise from 2010 to 2019. There is little production for African countries and a slight increase, and the differences are barely visible from 2010 to 2019.

8.3.2 Solar energy in euro-African countries (2010–19)

Table 8.2 shows solar energy production for African, European, and other countries of the world, starting in 2010, with 239 MW of African production capacity, twice as much for Europe as 30,856 MW and 41,542 MW for the rest of the world, while in 2013 the production capacity of African countries reached 724 MW and multiplied to European countries 84,261 MW and 139,582 MW, and 2017 saw some improvement in solar energy production for African countries of 4249 MW and 112,524 MW for European countries with 388,557 MW for the rest of the world. Still, by 2019, the production volume was 7443 MW for African countries and 140,587 MW for European countries with 586,434 MW for the rest of the world; through Table 8.2, we conclude that:

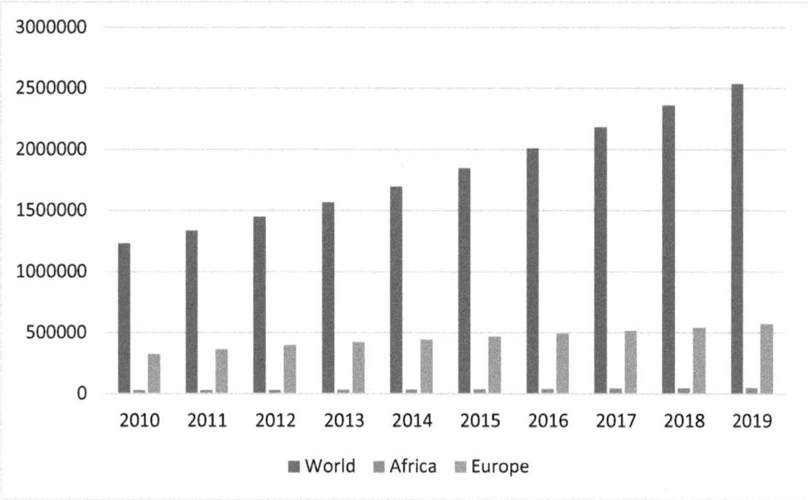

Fig. 8.1 The total renewable energy in Euro-African countries (2010–19).

Although the African continent can excel by exploiting solar energy by owning the Sahara, the world's largest hot desert, it is lagging in using solar power and solar thermal energy production.

This delay is due to the high solar cells' high prices imported from outside the continent at exorbitant prices, which is also expected to damage the solar cells after being exposed to the dust flying after the sand tornadoes. This delay is due to the absence of a strategy by African governments to exploit solar thermal energy, generate returns, and abandon fossil energy, even though it is inexhaustible energy and its returns are guaranteed—lack of private sector investment in renewable energies in general.

The solar panels acquired by African countries are not prepared and are not adapted to the African desert. Governments in African countries must open the door to the construction of solar-powered centers. The establishment of local research laboratories to open the way for research and development efforts to reach value-creating ideas that will find solutions to solar cells adapted to the African desert environment and which can do the self-cleaning process the importer cannot do so, encourage humans and experienced people to work locally in African countries, and provide a regulatory environment that helps create an atmosphere of creativity and create the right motivational methods for survival instead of migrating to Europe.

Fig. 8.2 shows high production of solar thermal energy for European countries and the rest of the world, where the curve is witnessing an upward trend from 2010 to 2019, while African countries are stable and moving in the same product and do not show any change from 2010 to 2019.

Table 8.2 Solar energy in Euro-African countries (2010–19).

CAP (MW)	2010	2011	2012	2013	2014	2015	2016	2017	2018	2019
World	41,542	73,734	104,078	139,582	176,018	221,993	295,821	388,557	488,752	586,434
Africa	239	331	411	724	1722	2248	3397	4249	6172	7443
Europe	30,856	54,719	73,801	84,261	91,125	99,592	106,094	112,524	121,517	140,587

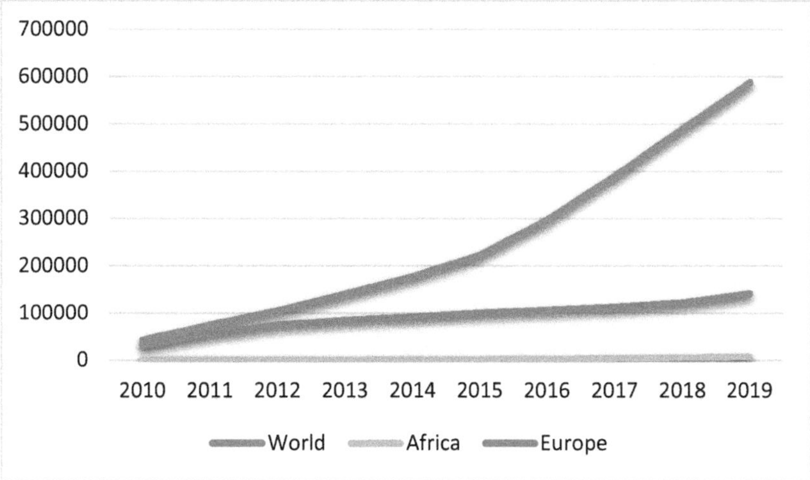

Fig. 8.2 Solar energy in Euro-African countries (2010–19).

8.3.3 Wind energy in euro-African countries (2010–19)

Table 8.3 shows the volume of wind power for African, European countries, and the rest of the world. In 2010, the report shows that the volume is 861 MW for African countries, 84,920 MW for European countries, and 180,850 MW for worldwide, While for 2019, the table indicates that the volume is 5765 MW for African countries, 195,908 MW for European countries, and 622,704 MW for worldwide. We conclude from the table the following.

European countries have substantial environmental potential to exploit wind energy; European countries have invested heavily in wind energy use to produce electricity and renewable electricity. They are establishing institutes and research centers that enable them to make the best use of wind energy and human resources and the care they receive.

Fig. 8.3 shows the size of wind energy, where the trend is an upward sway for European countries and the rest of the world, which occupy a significant share and their contribution significantly in the field of wind energy, in contrast to a curve that also climbs to African countries, but their contribution is hardly apparent, as these countries have not paid proper attention to the field of renewable energy and wind energy in particular despite the environmental potential of African countries, due to the lack of interest in the area at the expense of fossil energy.

Table 8.3 Wind energy in Euro-African countries (2010–19).

CAP (MW)	2010	2011	2012	2013	2014	2015	2016	2017	2018	2019
World	180,850	220,020	266,909	299,920	349,300	416,276	466,827	514,402	563,820	622,704
Africa	861	992	1125	1739	2398	3318	3830	4576	5465	5765
Europe	84,920	94,714	107,191	118,263	130,204	143,033	155,756	170,633	181,885	195,908

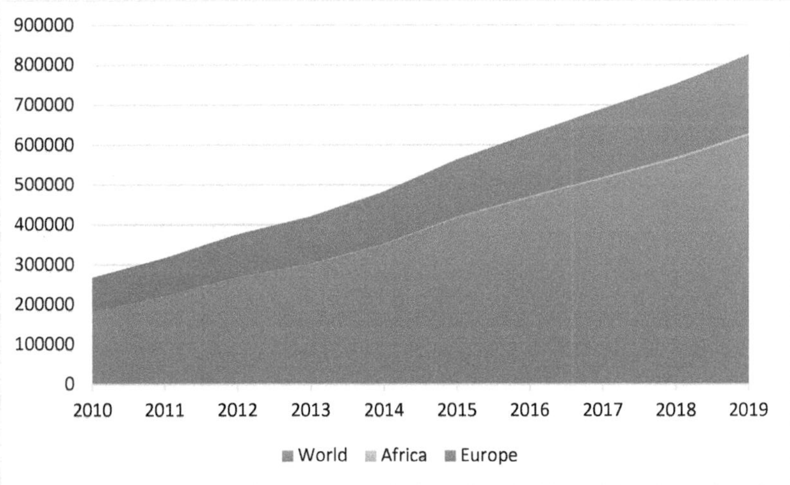

Fig. 8.3 Wind energy in Euro-African countries (2010–19).

8.3.4 Hydropower in euro-African countries (2010–19)

Table 8.4 and Fig. 8.4 show the volume of water capacity for African, European countries, and worldwide. In 2010, the volume reached 26,883 MW for African countries, twice as much for European countries 206,190 MW, and 1024,833 MW for worldwide. In 2013, the volume is 28,875 MW for African countries, 211,267 MW for European countries with 1135,510 for the rest of the world, and in the year 2017 with 34,845 MW for African countries and 219,690 MW for European countries with 1275,538 MW for the rest of the world, while its volume in 2019 was estimated at 35,887 MW for African countries, 221,802 MW for European countries, and 1310,292 MW for the rest of the world. Through the table, we conclude that the African countries have lagged behind the European countries and the rest of the world in the field of water energy exploitation, despite the environmental potential, due to the lack of material and moral support for those in charge of renewable energy in these countries, the lack of financial budget allocated to hydropower, and the absence of private and foreign investment, in contrast to the European countries that have mobilized the material and human organizational and organizational possibilities for the advancement of water energy and achieve returns and the results are impressive and translated into Fig. 8.4.

Table 8.4 Hydropower in Euro-African countries (2010–19).

CAP (MW)	2010	2011	2012	2013	2014	2015	2016	2017	2018	2019
World	1,024,833	1,055,557	1,088,563	1,135,510	1,174,396	1,210,496	1,245,469	1,275,538	1,297,465	1,310,292
Africa	26,883	27,021	27,680	28,875	28,965	29,490	31,414	34,845	35,691	35,887
Europe	206,190	207,950	209,347	211,267	211,993	214,537	218,114	219,690	221,191	221,802

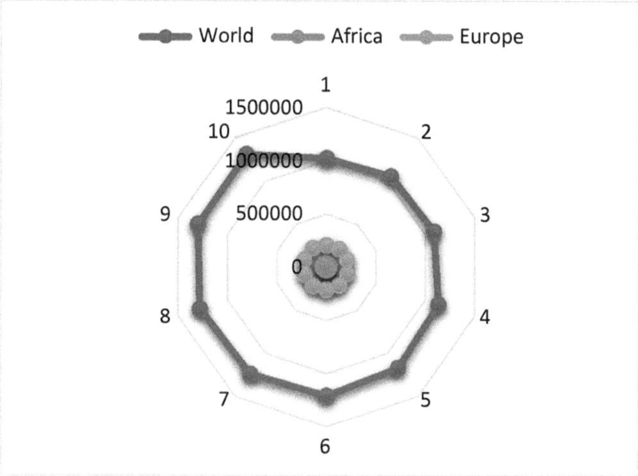

Fig. 8.4 Hydropower in Euro-African countries (2010–19).

8.3.5 Geothermal energy in euro-African countries (2010–19)

Table 8.5 shows the volume of ground energy for African and European countries and the rest of the world, where in 2010 for African countries it was 205 MW and 1336 MW for European countries with 9992 for the countries of the world. While in 2014, it was 373 MW for African countries and 1504 MW for European countries with 11,159 MW for the rest of the world, in 2017 it was estimated at 680 MW for Africa and 1556 MW for Europe with 12,700 for the rest of the world and in 2019 it was 830 MW for Africa and 1654 MW for Europe with 13,931 MW for the countries of the world.

Curve in Fig. 8.5 shows the volume of ground energy or ground energy, where the world's production is increasingly upward, which explains the world's interest in underground and renewable energies in general. European countries come second in terms of exploitation through a stable curve at the beginning and then upward from 2010 to 2019, while African countries are in the last place with an upward but very high curve, highlighting the weakness of exploitation despite the abundance of resources.

Table 8.5 Geothermal energy in Euro-African countries (2010–19).

CAP (MW)	2010	2011	2012	2013	2014	2015	2016	2017	2018	2019
World	9992	10,081	10,481	10,718	11,159	11,814	12,255	12,700	13,249	13,931
Africa	205	205	213	213	373	626	670	680	670	830
Europe	1336	1441	1450	1462	1504	1504	1504	1556	1615	1654

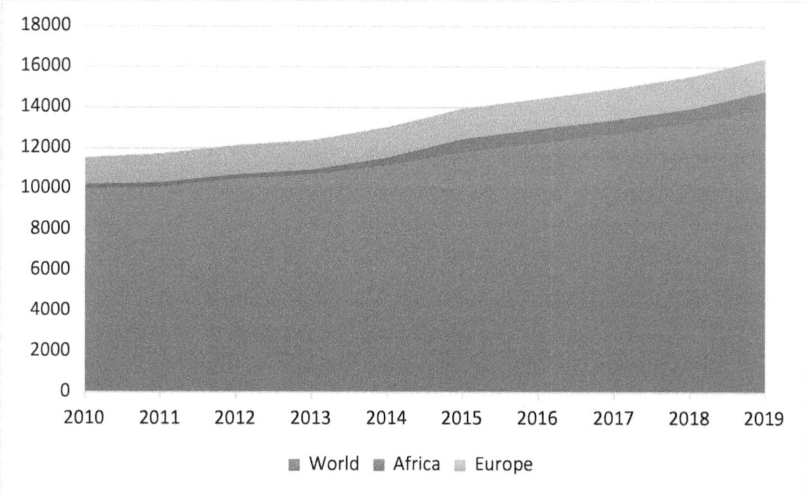

Fig. 8.5 Geothermal energy in Euro-African countries (2010–19).

8.4 Desertec project and the Algerian-German partnership on renewable energies

8.4.1 Introduction to the Desertec project

Desertec initiative is a win-win situation, with realistic and idealistic advantages (Slaoui, 2012). The Desertec Industrial Initiative was founded by several predominant German firms in 2009 (Schmitt, 2018). A solar project in North Africa was proposed by the Desertec Foundation of the Club of Rome, where the cost of the investment exceeds 400 billion euros and depends on solar thermal energy for the export of electricity produced from the project to all of Europe as well as for benefits of the countries of North Africa. The goal of the DESERTEC project is to achieve 15% of Europe's energy need conditions by the year 2050 (Dessouky, 2013). It should be noted that the project will rely on solar thermal energy and not solar cells on an area of 6500 miles or 17,000 km^2 in the Sahara desert to produce electricity from the European countries. Through a substantial high-voltage cable network, the project involves constructing a powerful connected system of mirrors to convert sunlight into thermal energy that heats a special oil used. The project could provide about 15% of the continent's electricity needs and is expected to get 20 gigawatts by 2020 and 100 gigawatts by 2050. Profitability from the other 19 different enterprises in the industrial initiative is justified because it is no longer comfortable with the commercial objectives expressed by the parties involved in the project, and general manager Paul von Son took over the leadership of the company alone, which stated that the project was focused 4 years ago on importing alternative energy from North African countries and now this method of thinking has been abandoned and the nonprofit enterprise of Desertec has been strongly supported by the German government, while the Desertec Industrial Initiative includes many multinational companies, but social and political unrest and the reluctance of the governments of North African countries have played an essential role in blocking the plans of the Desertec project, as it was recognized that Aruba could use Desertec as a means in its foreign policy to support the stability of North Africa when oil production declines.

In 2020, the Algerian government proposed reviving the idea of the Solar Desertec solar project more than 10 years after its first appearance to bring back to the minds of Algerians this project, which has sparked a lot of controversies. The Algerian Minister of Energy confirmed on the sidelines of the presentation of the government's work plan in the Algerian parliament to re-establish the project of electricity production by solar electricity desire, and said, "We have been consulting for 3 months with German companies to prepare an agreement allowing for a framework for talks and reestablishment of communication." Regarding the Desertec project, he added that it is expected to sign an agreement between the National Foundation Sounlgaz and German companies before next April. The goal is to complete solar panel plants in the Great South to achieve a healthy transition on high technological bases.

8.4.2 Objectives of Desertec

A few points can be included: In Algeria, energy usage is almost exclusively dependent on fossil fuels, hydrocarbons, and gases (Arabeche & Abdelli, 2020). The energy shift toward the abandonment of fossil energy, especially if we know that Algeria is a strong country of the largest producers and exporters of petrol and natural gas in the world and an essential member of the Organization of Petroleum Producers (OPEC), is expected to exploit shale gas, thus adopting the energy shift by the Algerian government and the orientation toward energies and the exploitation of renewable electricity is a recognition of the growing role of renewable energy in supporting the national economy. The redistribution of Algeria's electricity system in return will pay a nominal amount and end this project with power outages, especially in the summer, pressure times, and substantial electricity bills, exploiting and utilizing the most sustainable solar energy in the future and supporting the Algerian economy, promoting scientific research and research and development efforts in Algeria through the company with German-specialized laboratories and sharing experiences, And exporting renewable electricity and strengthening partnerships in the field of environmental technologies and industry with renewable energies and the conversion and dissemination of knowledge and technology. One of the project's objectives is to reach 5600 MW of renewable electricity production by 2027 with a vision for the future by 2050, which seeks to export renewable electricity to all parts of the world through a maximum production capacity of 7100 MW. The return on capital is invested for the Algerian state and the group of German institutions contributing to this project through technologies related to renewable energies.

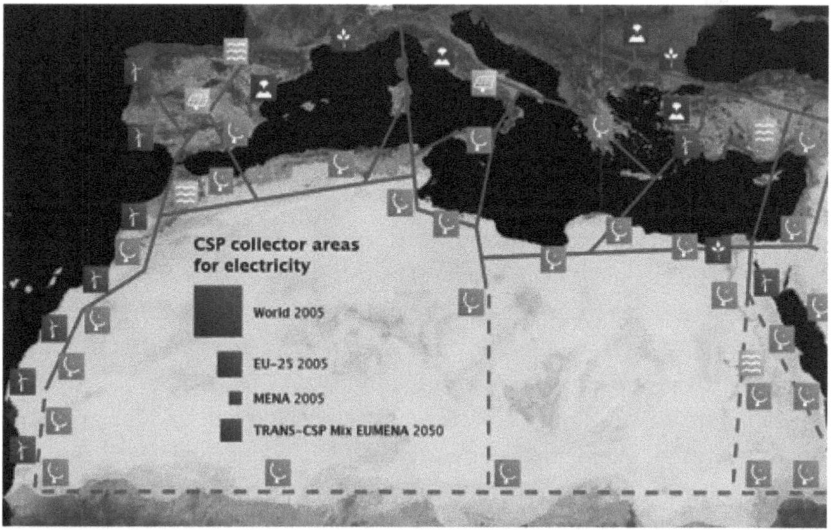

8.4.3 Reasons for partnership with Algeria

Located in North Africa, it is the largest African country and the tenth-largest country in the world in terms of area of 2381.741 million square kilometers, geographically from north to south 1900 km east-west 1200 and 1800 km off the coast.

8.4.4 The Algerian Sahara has the following characteristics

The Algerian desert is the largest solar reservoir in the world, according to German experts and the German space agency where solar radiation in the

Algerian desert lasts 4000 h of radiation per year and is the highest level of sunshine in the world according to the same experience and the production of electricity from solar energy can cover 50 times the needs of the European and African continent combined in the power of 1 year.

The Algerian Sahara is located in central North Africa. It is part of the Sahara Desert, representing 20% of the Sahara desert's total area and 80% of Algeria's total area. The Algerian desert has six hottest regions in the world and recorded temperatures such as the city of Bordj Badji Mokhtar at 60°C, Ain Qazam at 58°C, Ain Saleh at 55°C, Timimoun at 52°C, Adrar at 53°C, and Rujan at 59°C. The Algerian Sahara is the gateway to Africa's Sahara Desert, which spans more than 3.5 million square miles and is roughly the United States' size in its total length. The Algerian desert is a groundwater reservoir, abundant because of a thick layer of groundwater, and is mostly untapped. The Algerian desert is also the most extensive open garden in the world, known as the International Park Tassili n'Ajjer, which is a mountain range with waterfalls and a nature reserve, and the park has a unique ecosystem because of the amount of water available and the diversity of vegetables and animals in the heart of the Sahara. There is also a diversity in Algeria's climatic regions from the Mediterranean region in the north and along the coasts to the upper highlands region of the interior and then the desert region, which generally expresses desert areas interspersed with environmental and geographical diversity.

Algeria has access to private-public institutions interested in participating in the energy transformation process as well as institutional structures specializing in renewable energies, including C.D.E.R. that is tasked with collecting information and data to evaluate solar, wind, geothermal, and biomass energy, while formulating the research work needed to develop renewable energy production. U.D.E.S. Solar Equipment Development Unit is tasked with developing solar equipment and technologies and completing experimental models related to home, agricultural, or industrial, thermal effect solar installations, and solar installations due to voltage lighting and working to develop them according to usage. APRUE is established by the government to activate and implement the energy control policy and implement the programs approved in this framework. New Energy Algeria is a mixed company between The National Company Sonatrach, Solgas National Company, and Sim Food Industries Complex established in 2002, from its upgrading and developing renewable energies, completing projects related to renewable energies at home and abroad. There is interest from the Algerian state in the field of renewable energies through the announcement of future projects, including the task of the National Electronic Industries Corporation to complete a factory for the production of photovoltaic panels with a capacity of 670kW, 15% is directed to the local market expected by the year 2020. Cevital complex project in partnership with a foreign client to complete a 500MW photovoltaic production plant is underway. The Oras Solar project, in partnership with Algeria and France, to construct a 100MW annual PV plate production plant is underway.

8.5 Result & discussion

- Algeria has financial resources, environmental and regional diversity (coasts, desert, and fertile lands), and a large geographical area. Therefore, it surpasses Germany in terms of resources, mostly if it is known that Germany is a small country in terms of space with a moderate climate. There is no diversity in the regions; as for research centers, they are available in two countries at the highest level.
- Due to the weakness of environmental resources for Germany, especially about solar energy, as it witnesses solar radiation in a few periods of the year, but through scientific research and the exploitation of research and development efforts, it was able to reach the second largest producer of solar energy in the world for the year 2016. It is expected to fill 50% of its needs by the year 2050 to implement the Desertec initiative. Research universities have established 144 specializations in solar energy technologies, and the sector of business volume on environmental technologies is witnessing an unprecedented revolution in Germany.
- The partnership between European and African countries with environmental qualifications, such as Algeria, allows investment in renewable energies, as was the case in the Desertec project and the Algerian-German partnership.
- Renewable energy is considered inexhaustible energy, as green energy is called for its preservation of the environment. The most important thing is that it is easy to use and is not dangerous, which encouraged decision-makers to go into the field while minimizing hazardous nuclear energy and polluting carbon gas and other types.
- Investing in renewable energies came to address environmental problems such as emissions of harmful gases and industrial waste for developed countries until they later became profits and returns on invested capital.
- Investing in renewable energies depends on Europe and Africa's political will, which is responsible for evaluating the efforts made in this field

and the material and moral support of those in charge of this sector to embody the future strategic vision of initiatives such as the Desertec initiative.

- Investing in renewable energies also depends on the legal regulations in this field, as it governs the method of exploitation, the place of exploitation, and those in charge of exploitation in addition to the responsibilities, powers, and accountability, and these legal controls are characterized by flexibility as they are subject to constant amendment and in line with modern changes.
- Investing in renewable energies relied on stimulation, whether for institutions by providing support, facilities, or for individuals through reductions in the bills and pricing of renewable electricity, for example, which encouraged members of society to move toward renewable energy consumption without the energy that comes from fossils.
- The media propaganda for renewable energy played an important role in mobilizing public opinion about its positives with its advantages by creating periodicals concerned with this topic. Information was presented with statistics on the subject and the launch of initiatives such as Desertec.
- Investing in renewable energies between the European and African continents has contributed to exchanging experiences and technologies to preserve the environment and various experiences and equipment.
- The private sector's participation in investments in renewable energies in Germany contributed to reducing costs while obtaining deducted returns from the properties and achieving competitiveness in this field.
- Investing in renewable energies has made it possible to open significant job positions and reduce unemployment, especially if we know that this investment involves government and private sector institutions.
- The German experience should be adopted as a successful model to be emulated in developing countries that are still behind in this field, despite the great potential of most of these countries. The Algerian experience is the best proof, and thanks to the geographical and environmental capabilities, the results are meager, but cooperation with Germany has a positive effect.

8.6 Conclusion

The Desertec initiative has not achieved its goals 11 years after its launch, due to financial, legal, and administrative obstacles. Our study focused on this strategic project that aims to provide European countries with renewable energy from Africa's deserts. Still, the project was launched in Morocco and did not start in several countries such as Algeria, including technical, economic, and legislative.

Among the recommendations that the research concluded are:

- Severe methods must be found in research and development in renewable energies to reach more advanced technologies that will utilize renewable energies.
- European countries should utilize the available resources to improve statistics and achieve more returns on the capital invested in renewable energies.
- The development process in environmental technologies requires excellent efforts, and an exchange of experiences is necessary between Germany and European and African countries to achieve sustainable development.
- Modern technology and equipment should also be transferred to African countries such as Algeria, which are still lagging in this sector. Renewable energy projects can also be exchanged between the European continent and Africa, as we studied in our chapter on the Desertec project to achieve economic growth.
- The renewable energy industry is an emerging sector in African countries. Experts from developing countries have not played an influential role in its development. Therefore, the renewable energy strategy should be planned with economic actors, stakeholders, and experts to research this vital industry and analyze strengths, weaknesses, and opportunities to produce results for preparing a national strategy in renewable energies.
- The trend of economic policy in Algeria toward an interest in the field of renewable energy, based on the enormous potential of an area of more than 2.3 million square kilometers of solar radiation throughout the year and significant wind energy in the sea and on land, and in light of the depletion of fossil energy and the economic austerity crisis, the energy field remains. Renewable is the ideal solution for achieving significant economic growth.
- Legislation in renewable energy and activating mechanisms to define methods of work, exploitation, and responsibilities.
- Establishing a national observatory of data on Algeria's geography shows the areas for installing renewable energy installations, and the process should be based on scientific measures.

- The renewable energy industry is a way to solve the environmental problem. Still, in Algeria, there is a problem of groundwater rising to the earth's surface in the desert, and this opens another way for the exploitation of renewable underground energy in southern Algeria, especially when the water resources are scarce in that region.
- A renewable energy economy is considered a practical solution to eliminate unemployment by absorbing many individuals in Africa's private and public sectors, especially Algeria.
- It establishes strategic projects that serve the leading countries in Europe, such as Germany, and developing countries such as Algeria in the renewable energy sector.

References

Arabeche, Z., & Abdelli, M. E. A. (2020). Insight of green economy in Algeria. In *Eco-friendly energy processes and technologies for achieving sustainable development* (pp. 108–126). IGI Global.

Dessouky, M. O. (2013). The environmental impact of large scale solar energy projects on the MENA deserts: Best practices for the DESERTEC initiative. *Eurocon, 2013*, 784–788.

Harjanne, A., & Korhonen, J. M. (2019). Abandoning the concept of renewable energy. *Energy Policy, 127*, 330–340.

Kabir, E., Kumar, P., Kumar, S., Adelodun, A. A., & Kim, K.-H. (2018). Solar energy: Potential and future prospects. *Renewable and Sustainable Energy Reviews, 82*, 894–900.

Kaya, D. (2006). Renewable energy policies in Turkey. *Renewable and Sustainable Energy Reviews, 10*(2), 152–163.

Lowitzsch, J., Hoicka, C. E., & Van Tulder, F. J. (2020). Renewable energy communities under the 2019 European clean energy package–governance model for the energy clusters of the future? *Renewable and Sustainable Energy Reviews, 122*, 109489.

Moran, E. F., Lopez, M. C., Moore, N., Müller, N., & Hyndman, D. W. (2018). Sustainable hydropower in the 21st century. *Proceedings of the National Academy of Sciences of the United States of America, 115*(47), 11891–11898.

Nelson, V., & Starcher, K. (2018). *Wind energy: Renewable energy and the environment*. CRC Press.

Pacesila, M., Burcea, S. G., & Colesca, S. E. (2016). Analysis of renewable energies in European Union. *Renewable and Sustainable Energy Reviews, 56*, 156–170.

Schmitt, T. M. (2018). (why) did Desertec fail? An interim analysis of a large-scale renewable energy infrastructure project from a social studies of technology perspective. *Local Environment, 23*(7), 747–776.

Slaoui, A. (2012). Editorial: Desertec project—When science joins politics. *Journal of Renewable and Sustainable Energy, 4*(1), 10401. https://doi.org/10.1063/1.3687000.

Toklu, E. (2017). Biomass energy potential and utilization in Turkey. *Renewable Energy, 107*, 235–244. https://doi.org/10.1016/j.renene.2017.02.008.

Wüstenhagen, R., Wolsink, M., & Bürer, M. J. (2007). Social acceptance of renewable energy innovation: An introduction to the concept. *Energy Policy, 35*(5), 2683–2691.

CHAPTER 9

Connecting ecological economics, green management, sustainable development, and circular economy: Corporate social responsibility as the synthetic vector

Diana Fernandes[a] and Carolina Machado[b]
[a]University of Minho, Braga, Portugal
[b]School of Economics and Management, University of Minho, Braga, Portugal

9.1 Introduction

As of the 1990s, globalization has been decisively impacting the global order, reconfiguring it to a new era marked by the redefinition of the world as a global village, a metaphor which is frequently invoked to portray the current complex global heterogeneity, and the increasing interaction among several actors despite time and geographical limitations, thus climaxing into the deep interconnectedness and integration of economic activities of different human societies around the world (Ajayi, 2001). In other words, nonetheless in concordance to the above, globalization has been the defining trend on the worldwide balance of power ever since the closing decade of the 20th century and the dawn of the new millennium, signaling a new global era evidently marked by the deep, complex, and wide-ranging interaction among nations, economies, and people, via a transversal and ongoing process of global integration of the economic, financial, political, and cultural fields.

Deriving from the above, we postulate that, in concrete terms, globalization crystalizes the picture of an evidently borderless world accelerated by increasingly greater economic integration, which (theoretically) is presupposed to enhance the living standards of people across the globe, fostering

Green Production Engineering and Management
https://doi.org/10.1016/B978-0-12-821238-7.00001-4

well-being at a global scale. Hence, it consists of the intensification of cross-border trade and increased financial and foreign direct investment flows among nations, both being encouraged and materialized by the rapid acceleration on the liberalization of communication, hence also simultaneously deriving and benefiting from the information technology (Islam, 1999).

The truth is, indeed, that the debate is twofold. From one hand, if acceptable argumentation can be done defending that the economic globalization reconfigures the government-market interface, constraining the governments' potential and scope for action by privileging the market dynamics as a core means to foster economic, social, and environmental outcomes. On the other hand, it is also reasonable to advocate that it generates new-fangled imperatives so that national states cooperate both in managing the global commodities and in coordinating domestic environmental policies, thus launching an effective collective effort directed for the common good, aimed at the overall societal well-being.

Economic globalization contributes to economic growth, nevertheless, and via a plethora of ways and a variety of channels, it as well impacts the environment and, subsequently, the potential for sustainable development. Namely, its dynamics sharply accelerate structural change, altering the industrial structure of countries and/or regions, which, afterward, generate complex challenges around resources' stocks. Acknowledging the fact that it conveys and amplifies market and policy failures and the respective impacts, it becomes obvious that its dynamics may spread and exacerbate environmental damage and the potential for survival not only for present generations, but also in regard to the upcoming ones (Gallagher, 2009).

Therefore, it may engender pressures toward profound reforms, given the fact that, progressively, strategies and policies hitherto conceived and implemented as purely domestic start to evidently attract international interest, due to their spillover effect beyond national and, even, regional scopes. So, from this an assumption derives: while globalization improves and expands the projections for economic growth worldwide, as by increasing overall global output, it may, as well, result in negative impacts, observed the fact that it plausibly reduces economic prospects in individual countries, sectors, and industries, accentuating regional dissymmetries in a myriad of scopes (economic, financial, political, cultural, etc.). Subsequently, it originates an increasing heterogeneity on the development potential, harming sustainability, as it produces dynamics of marginalization in regard to economies and people, which may be outcome in poverty-induced resource depletion and environmental degradation (Destek, 2020).

Following this line, Roome (1992, p. 11) indeed alerts toward the "critical turning point for society, business and environment," tightly visible ever since the 1990s, an idea corroborated by Vlados, Deniozos, and Chatzinikolaou (2018, p. 21) when claiming that the "idiosyncratic track of the current crisis of Globalization" has been plainly and directly imprinted "in the deepest level of the socio-economic evolution of every region and every agent of action on the planet," so that it "penetrates and dissects the real life of every individual, group and the entire organization, their plans and actions of every kind."

So, in the contemporary context, organizations can by no means overlap and ignore environmental issues, acknowledging the assumption that all dimensions of globalization affect the natural environment and, through it and via a spillover effect, also human development. Indeed, in the current era, we witness increasing government regulation and growing scrutiny condensed on a loud pressure for stronger public mandates toward environmental accountability (Walton, Handfield, & Melnyk, 1998).

Following this, the *milieu* to the choices society faces in regard to the environment has been set out in the literature as analyzing the difficult and traditionally conflicting (this to say, mutually exclusive) connection between conservation and development, expressing that society needs to urgently reshape the economic, financial, political, and social paradigms in order to effectively move toward environmentally sustainable approaches, which require an ultimate reassessment of our strategies, policies, activities, institutions, and organizations. Hence, there are growing calls toward the need for society to acknowledge that its current lifestyles plainly contribute to environmental change on a significant, and indeed potentially life-threatening, scale (Fernandes & Machado, 2020).

Therefore, it is urgent that academic, political, and economic debates agree on the acceptance of an environmental ethic as a pillar to underpin individual and social behavior. In other words, the refined approaches fully recognize the need for a radical shift in how society views and understands the world's resources as an imperative to face the challenges crystalized in this new era brought about by globalization.

This is expanded when considering effective environmental decision-making, given the fact that proactive environmental management initiatives entails inherently multiobjective processes, which require joint considerations of environmental, industrial, economic, and social criteria in all stages of the decision-making process (Khalili & Duecker, 2013). This is to say that such process is extremely complex due to the systems' interdependent and complex dynamics, and to the competing interests of multiple stakeholders. This way an approach which actively addresses and manages them requires a transparent decision-making process, flexible to changing circumstances, and inclusive on heterogeneity.

Literature on environmental management advocates that, being organizations one of the main actors responsible to cause several of the current environmental problems, they shall as well assume an active and wide-ranging role not only in creating awareness on such issues, but also (and foremost) in effectively acting toward them. Subsequently, we assist to a growing set of eco-initiatives launched by organizations to address environmental management concerns, as one concerted organizational response to environmental degradation, supported on a green management model.

So, with this essay, we seek to weave a theoretical and conceptual reflection around the influence of social, economic, financial, political, environmental, and other external forces on the current economic approaches directed at the ecological domain, in order to suggest a multidisciplinary logic of reasoning as the most updated and accurate approach in order to understand the current global economic, political, and social dynamics, based on the distinctive challenges brought about by the new worldwide order generated by the impacts of globalization.

We intend to launch a discussion and develop it to a more refined level of analysis, around how the current global order is marked by the vector of change, and so, how the current worldwide dynamics are increasingly challenged by the growing acceleration and diversity, questioning their potential to indeed impel the world toward the progress on wide-ranging development, directed at the maximization of societal well-being. Hence, the analysis we do is fueled by the insights of several disciplines, interconnecting economic, management, ecological, political, and even sociological insights. It aims to understand the core evolutions on economic thinking in which regard to the natural resources as a vital asset for economic growth, and how the threats along the natural environment have harmed their exploitation and conservation potential to levels in which the conventional neoclassical economics needed to rethink the classic postulates in order to review the dyadic relationship between human beings and nature as complementarity (via win-win games) and not as a mere utilitarian exploitation (via zero-sum games).

Hence, we expose the progress of the economic discourse, progressively acknowledging the challenges of environmental management to be included as a crucial premise on the modern economic thinking, which then realized the need to imprint economics as a true social science, vesting it on its social input conventionally neglected in neoclassical approaches. This way evolving via more transversal and multidisciplinary approaches, it then reached a refinement level in which it clearly directs toward sustainability, in order to contribute to the improvement of the social and environmental performance of organizations, adding value of social relevance, calling out for a green management frame, and acknowledging the articulation of the worldwide dynamics as a deep and complex network of several stakeholders and interests, condensed best via the postulates of a circular economy. Therefore, we intend to focus on strategies, policies, and practices that support the paradigm shift in organizational and people management around environmental issues, considering people as the core asset for both economic growth and global development, assuming the overall societal well-being as such approaches' twofold key premise and goal, and preserving the environment as a means for maximizing the present generations' capacity to satisfy their needs, without compromising future ones to meet their own demands.

In methodological terms, this chapter presents a descriptive work based on reviewing literature and empirical data of the most relevant contributions in the topic, and thus, it assumes a fundamentally theoretical nature.

There are several contributions provided by this work, and by presenting them, we can as well dissect the chapter's structure. First, we present the core conceptual triangle formed by the analysis of the conceptual constructs of *Globalization*, *Social Pollution,* and *Change*, dissecting its interconnectedness in order to ascertain the current global scenario and its challenges which fundament the urgent need to review economic, political, and social approaches, framing them on the motto for sustainability. Based on this contextual exposition, the chapter then surveys and draws together the elements of environmental management, acknowledging that one of the main challenges brought about by globalization—both for the potential for economic growth and for the global societal development—consists of the natural environment conservation, due, for example, to the growing shortage of natural resources stocks, pollution problems, etc. In the third section, the chapter furthers the above insights in order to highlight the relevance of sustainable management as the conceptual trigger to raise the *momentum* for green management. Hence, this section aims at exposing the main premises and managerial approaches advocated by the green management, highlighting them as an evolution from the presupposes of sustainable management, hence as well describing how they arise as a progress of the concerns already addressed by early works on the ecological economics domain.

Thus, all the argumentation to that point constructed is then taken to a refined level of analysis, by shedding light on the theoretical and managerial approach condensed around circular economy, advocating that it marks a prime approach to conceive the current global dynamics and its respective challenges, as it indeed allows a fusion of economic growth and societal development premises, articulating them into a joint commitment, via a symbiosis dynamics aimed at reaching an equilibrium toward the achievement of particular business interests and the maximization, as well, of the public good—meaning, the societal overall well-being. Hence, this section subdivides into four subsections, namely a first intended at highlighting people as the core asset for societal development and sustainability; a second aimed at connecting ecological economics, green management, sustainable development, and circular economy by suggesting corporate social responsibility (CSR) as the synthetic vector along all of the abovementioned constructs; a third which alerts for the calling challenge toward a synergetic political and economic discourse, progressing from welfare postulates to the well-being ones; and finally, a fourth summarizes all of the inputs provided in order to actively fundament the societal overall well-being as the nuclear (and hybrid) input and output advocated by the approach defended all along with the chapter.

Conclusions, limitations, and avenues for future research, then, follow.

Henceforth, the relevance of this chapter would be responding to calls in the literature in order to integrate ecological economics, green management, sustainable development, and circular economy as an urgent and interesting subject of research, via concerted, transversal, and multidisciplinary approaches, both in the academic, political, and managerial domains (Jabbour & Santos, 2008; Jackson, Renwick, Jabbour, & Muller-Camen, 2011), and to expand the scope of conventional economic thinking, primarily grounded under a strategic management paradigm, to indeed incorporate sustainability as a core premise. This way, taking active steps to be framed under a Sustainable Management context by acknowledging the calling environmental challenges, firmly assuming people as the core asset for the world development. And understand economic growth not by an end per se, but as a means to effectively affect the worldwide societal well-being (Osland & Osland, 2007; Wilkinson, Hill, & Gollan, 2001).

From the present chapter, we can conclude that for today's organizations it is vital to promote a proper alignment of resource management principles with green management, so that the organization itself fully embraces the dynamics of a circular economy, hence considering itself as the joint strategies, policies, and actions of a myriad of stakeholders, via a constant commitment, assuming an active role on the global development, because its actions spill over traditional border of business sector, market, geography, and even time. We, thereby, conclude that the greater the strength of green, sustainable, and circular strategies, policies, and practices, the greater the organizations' intensity to adopt environmental management systems and policies will be, under the premise of fostering overall societal well-being (prosperity) beyond economic growth, inherently referring to sustainability as an imperative.

Nonetheless, this work presents limitations as it consists of a pure theoretical effort, thus still lacks empirical corroboration. We also verified some semantic and managerial juxtaposition among several of the conceptual constructs on which the chapter is built; hence, we suggest that these needs to be first properly addressed and clarified.

We, thus, suggest that future research shall, undeniably, rely on multidisciplinary, transversal, and holistic approaches. The sustainable competitive advantage would be an interesting avenue for the forthcoming works, and expanding this idea, we also recommend that Green Human Resources Management assumes considerable potential as a research area. We can note that the path to the implementation of a Green Human Resources Management paradigm lies on the setup of supportive cultures for environmental management, this based on the fact that effective outcomes in terms of environmental management are achieved not just by making changes in production processes, products, or raw materials, but rather by fully adapting the corporate culture so that we can have organizations profoundly embedded toward values effectively supportive of long-term sustainability (Kitazawa & Sarkis, 2000; Stone, 2000). So, identifying sources of both resource capital and institutional capital among competitors may shed additional light in order to foster the organizational sustainable competitive advantage, hence stressing the relevance of longitudinal studies on resources' development.

9.2 Globalization, social pollution, and change: The core conceptual triangle

Capitalism has been solidified as the model of the current economic and financial worldwide order (and, given its consequences, also political and social). Situation which has been boosted by the dynamics of the economic and financial globalization, being Capitalism considered as the vehicle for worldwide economic growth and societal development. The corollary of this situation conveys, as a thesis, that organizations which stimulate profit will merit privileged treatment, their policies, and practices will be conceived as correct, operationalizing efficiency. So, they will be set free of restrictions in their policies and practices, so that they can maximize performance, this given the fact that the configurations we are experiencing tend to reflect the idea that efficiency will be achieved by abolishing restrictions on markets, from which it follows that government intervention or regulation of markets, including labor markets, will be counterproductive and harmful (Pfeffer, 1998, referred by Fernandes and Machado, 2020).

Thus, recent decades have witnessed the intensification of international capitalism—these, according to Deluiz (2017), can be listed as some of the signaling elements of the structural transformations, which indelibly mark the contemporary context and impacts on the future.

Undeniably, globalization "can be conceived to be born and reproduced by the diverse social and economic interconnections and flows that it creates and exploits," based on the premise that a variety of complex dynamics (economic, financial, social, political, institutional, demographic, cultural, religious, climatic, geopolitical, and geostrategic) are mingled and redefined to a notable level (Avant, Finnemore, & Sell, 2010, p. 10).

It is indeed in this context that we can call out the concept of "Social Pollution" as clear social negative impact from the accelerated and heterogenic economic and financial dynamics bought about by the globalization, alluding to the argumentation of Pfeffer (2010).

Such practices have serious harmful effects not only on the physical and psychological well-being of people (Price, 2006), but also on the communities where they work and cohabitate (Pfeffer, 2010). Despite the current recognition of this situation, the adoption of measures to prevent its harmful effects still falls short of expectations (Guthrie & Datta, 2008).

Hence, it is possible to advocate the invisibility of the societal consequences that harmful people management policies and practices can unfold. This is so because, although organizations know what to do to make their workplaces more efficient and sustainable, relatively few actually commit to concrete actions, which highlights the urgency and complexity of bringing the social effects of organizational management practices into the discussion. It is, therefore, important to understand why inefficient management practices persist, because the end of the "Social Pollution" (Pfeffer, 2010) created by toxic work practices makes socioeconomic sense not only for society, but also for the organizations themselves. According to Pfeffer (2010), a possible reason for the persistence of such paradoxical practices—and the related externalities—may root in low (and ineffective) social sanctions. In addition, it is important for organizations to be aware that there is no need for unrealizable compromises between doing well financially and doing good for the workforce and for the society.

Today, we act toward a fluid global environment, marked by constant and increasing heterogeneity, liquidity, and uncertainty (Vlados et al., 2018). Therefore, globalization and its interdepended complex worldwide dynamics are forcing agents to dramatic adjustments and improvements, in an increasing speed. For these reasons, Deluiz (2017) systematizes that the globalization of work and of the social question corresponds intrinsically to the economic globalization. Deriving from the same logics, the arguments of Pfeffer and DeVoe (2012) and of Vohs, Mead, and Goode (2008) allow us to verify the current prominence of conventional economic assumptions, emphasizing the belief in markets as the core agent for a decision-making logic which emphasizes individual needs over social ones—thus, competition over cooperation, leading to the neglect of social consequences in the face of economic ones. And this, following Bento (2011), propels negative outcomes, given the fact that in many societies a collective way of life has been established based on a level of well-being higher than the contemporary resources available and temporarily sustained, hence mortgaging the future well-being, which imposes a forced readjustment of the intergenerational contract. From this framework, the models of economic growth and development have been shaken.

As a result of the globalization and its imprint, the contemporary world is paved with increasing uncertainty and heterogeneity, which requires organizations to make changes (Chiang, 2010). These come at two levels. First, fundamental change requires changing culture and/or identity, based on the premise that sticking to an identity and denying the upcoming transformational forces may lead to failure in creating added value. Second, capacity for change implies that an agent shall be capable of responding quickly to what is occurring around.

It is in this sense that a purely economic/financial framework leads to the evaluation of social arrangements (only) in those same terms, a parsimonious and limited perspective, possibly counterproductive (Ferraro, Pfeffer, & Sutton, 2005). Based on this argument, *Change* emerges as a neuralgic concept to understand not only the advent of new dynamics, but also its implications.

The concept undoubtedly comprises a strategic dimension, as it refers to a vital progress which entails the movement of an agent from its current situation to a desired future. From this, we can derive the premise that adapting to change is the absolute key to survival: so, organizations (in the means of the core societal structures, thus defined as consistent and structured conglomerates of people) should prepare for the constant creation and recreation of complexity, in order to cope with the environment's itself broader complexity, a *sine qua non* condition in today's rapidly changing environment (Lorenzi & Riley, 2003).

From this line of thinking results the premise that managing change has plainly and sharply become essential as the ultimate component aimed at a successfully managing strategy in the contemporary society. Nevertheless, and not disregarding that win-win situations are possible in some societal agents and contexts, we assume that for the majority of them it will not be so easy to create simultaneously economic and financial efficiency alongside to ecological, social, and human sustainability without a core change in their strategy and culture (Ehnert & Harry, 2012). In fact, knowing the pace and unpredictability of change does not necessarily imply that agents have learned to manage change and that actively and efficiently do it: the core assumption here is that change redefines risk; henceforth, this premise infers that agility in adaption becomes more important than accuracy in reducing risk.

Consequently, the environmental challenge is growing awareness in all agents' daily routines, being sustainability the motto for today's world—we indeed assist to the increasing triumph of the "Green" mindset, which gradually calls out for the need to rethink the traditional pillars of the capitalist order. In the current frame of heightened scrutiny of corporate behavior (Basu & Palazzo, 2008; Waddock, 2000) and increasing demand for corporate social responsibility programs (Lenssen et al., 2009), societal agents (straight away organizations, as a collective structured aggregation of people, thus core of the societal order) are being progressively called out both as the main cause of environmental problems and as the main agents to raise awareness on the topic and to mobilize and implement actions aimed at effective solutions.

9.3 Rethinking neoclassical economics—Furthering the debate with ecological economics

From this background, we can then see that ecology and economics have traditionally been separated research fields, addressing dissimilar problematics, applying different assumptions, being developed under diverse conceptual structures, and supporting different interests in the policy-making process. Nonetheless, the advent of ecological economics marks a distinctive cornerstone, as it has reconciled both approaches, driving academical research, civil society debate, and political agendas toward a new paradigm (Fernandes & Machado, 2021).

It is imperious to quote Boulding, Daly, and Georgescu-Roegen as the intellectual founders of ecological economics, on the field of economics, as well as Holling and Odum on the ecology domain (for a consolidated and detailed review on the literature on this topic, please consult Fernandes & Machado, 2021).

In a simple motto, we can then see that ecological economists focus more incisively on the problems facing the *Homo sapiens*' (future) existence and well-being, as well as the ecosystems on which civilization rests on the long term. Hence, this advanced approach opposes to a strictly utilitarian conception of value (Sagoff, 1995), because it rebuts the mainstream perspective by arguing that "the basic relation of man-made and natural capital is one of complementarity, not substitutability" (Daly, 1994, p. 26). So, it embodies an holistic approach playing "a leading role in recasting the scope and method of economic science" (Gowdy & Erickson, 2005, p. 207).

Developing on the assumption that particular processes in nature are essentially irreversible (Clark & Munro, 1975; Costanza, 1996; Costanza & Cornwell, 1992; Ludwig, Hilborn, & Walters, 1993), the models brought about by the ecological economics suggest the need to leverage natural capital (Costanza & Daly, 1992) in order to preserve and reinforce the ecological life-support systems, as well as the linked socioeconomic systems, so that they remain resilient to change and uncertainty (Hammer, Jansson, & Jansson, 1993; Holling, 1994; Perrings, 1998).

So, the term consists of a profound transdisciplinary effort to broadly shed light on the connection between the natural and the social sciences, actively criticizing the blind isolation traditionally devoted to them (Costanza, 1991). In fact, ecological economists argue that mainstream economics—grounded on unidimensional dimensions such as complete measurability, value neutrality, and objectivity—consists of a reductionist approach, mainly focused on the processes of price formation, thus, being fallible and even incorrect in order to understand the current world dynamics, given the fact that it overlooks the systems' open and evolutionary nature. It conceives the economy as a purely closed system in which firms sell goods and services, and then remunerate the respective production factors. This way, conventional approaches neglect the distinguishing nonlinear relationships observed all along the economic system (Christensen, 1989), wrongly assuming that it is possible to cope complexity by dissecting the world into relatively isolated units, examining them on their own singularity, and later reassembling all of those in order to provide an overall picture of the initial complex context (Costanza, 1996).

Indeed, mainstream economists defend that it is not possible to scientifically access the Earth's carrying capacity, once it is subject to bias giving the fact that it consists of a mere result of the state of knowledge and technology at a present space and time context. So, the theoretical and empirical models postulated by such perspective root on trusting to human intelligence and ingenuity, postulating that, as people seek to meet their needs and preferences in order to achieve well-being, they act within the environment by choosing among a medley of alternatives (Barnett & Morse, 2013). Consequently, nature imposes no limits to economic growth, and even if such constrains exits, knowledge (assuming as the limiting factor in economic production) is expected to always lessen resource shortages (Drucker, 1993; Lee, 1993).

Exposing an energic contrast, ecological economists reject such premises on the basis of a different understanding of the limiting factor, advocating that it no longer refers to human created capital; instead, it consists of the remaining natural capital (Costanza, 1991). Personifying it on Daly's argument (1985), ecological economics strongly defend that economic growth faces (and indeed has to face) limits and obstacles, because the sources of raw materials (natural capital) are fixed and limited, restraining the global economy's growth potential.

Hence, and alluding to Rapport (1993) referred by Fernandes and Machado (2021), we can conceive ecological economics as both an anthropocentric domain (because, at its nucleus, it focuses on the survival and well-being of the human species as an active agent in a given environmental context) and a biocentric domain of research (because it also provides insights on the survival and well-being of all other forms of life). So, it does not extinguish on the *Homo sapiens* as the object of study, given the fact that it indeed consists of an effort to provide a more comprehensive science of humans as an active component of nature (Costanza, 1996).

As a synthesis, the argumentation so far exposed climaxes on the premise that the heart of ecological economics associates with the goal of sustainable management, which, by its own end, associates with the core values and aim of sustainable development, once all concepts interpret the economy as only a subsystem which is an active part of a larger local and global ecosystem, the latest decisively imposing limits to the global economy's physical growth. This argument grounds on the two fundamental questions addressed by ecological economics: the issue of environmental externalities and the optimal intergenerational allocation of nonrenewable resources. Hence, the ultimate goal is to design and implement management and political strategies, policies, and practices aimed at achieving both intra- and intergenerational equity, welfare, and well-being (van den Bergh, 2000).

9.4 Sustainable management and the *momentum* for green management

Strategic management has emerged as the dominant model in the literature in the economic and financial field in the last three decades; however, foremost as of the last decade we have been witnessing the solidification of a new approach, sustainable management. This new concept is based on the urgency of articulating an agenda for global change and the development of a common future for humanity, concerned with the interconnection of social and economic progress without threatening current and future living conditions. Thus, this new theoretical reconfiguration marks a shift on the conceptual paradigm, which coined a more vibrant social imprint on the classical economical and financial approaches—often deviated from the social agents and its dynamics.

Literature evidently allows us to understand the increasing and recent interest which has been paid to environmentalism globally. As the current trend, it is generally acknowledged that protecting, improving, and managing environmental resources are crucial. Therefore, this concept articulates an approach in which organizations emerge as prospectors and not only reactors in light of the current international context's volatility.

So, and under this theoretical frame, literature allows as to summarize that societal agents in the current worldwide spaces and dynamics will have to create value in three areas, deemed the three pillars of sustainability—economic, social, and environmental—going beyond the action within geographical, political, cultural, financial, economic, institutional, and organizational boundaries (Elkington, 2006).

In fact, organizations (as core people structured conglomerations, so core societal basis) increasingly need to consider themselves as performing a major environmental role, for example, aimed at ecological footprint reduction, and, thus, are pressured to ascertain the growing importance of social and environmental vectors along with economical and financial ones, so that they can be enabled to succeed not only on the corporate sector, but also in the global societal development (Cherian & Jacob, 2012). Indeed, a large number of organizations, institutions, and stakeholders (frequently with competing or even irreconcilable objectives) are responsible for policy analysis, regulatory decision-making, and priority setting for environmental actions and assessments, having a vibrant voice in the outcomes for global development and sustainability. As a corollary, organizational sustainability can be understood as the managerial practice that aims at timeless organizational vitality, performing ethically plausible actions that favor equitable economic, environmental, and social performance criteria (Jabbour & Santos, 2008).

As a result, current research presents an intensification (both on the quantity and on the complexity of studies) around the development and use of formal scientific approaches to assist in environmental management decision-making (Renwick, Redman, & Maguire, 2012), relying on sustainability as its core vector. This resulted in the advent of a new strategic theoretical and managerial maneuver called green management.

According to Munda (1997), the apprehension around the carrying capacity of Earth symbolizes a key withdrawal from traditional arguments in favor of environmental protection. In fact, vividly criticizing the neoclassical economists framed on a mindset which conceived nature as full of divinity, the author openly pointed the fragilities of such approach, once it did not rest on prudential considerations, thus, looked upon environmental conservation less as an economic imperative than as a moral test, disregarding the need for natural resources conservation due to their limited stock in nature and, in some cases, to their substitutability impossibility. Hence, as a vibrant contrast to such trend on the economic thinking and managerial analysis, ecological economics anticipated an explicit apprehension for future generations' ability to meet their needs, articulating their postulates around a clear focus on long-term sustainability and exploring questions around unconventional economic topics such as ethics, human rights, nature conservation, equity, regional development, and multiculturalism (Berkes, Colding, & Folke, 2000). Therefore, this renewed trend on the economic debate uttered through models entailing a wide range of values, via an harmonious dynamics in order to reconcile traditionally irreconcilable interests, providing a more ambitious and rigorous level of study as such models go beyond the fragmented insights of the current human generation—although these are not ignored (Fernandes & Machado, 2020, 2021).

In light of the above, and grounding on ecological economics as its theoretical foundation, back in the end of the 18th century we can find the roots of the concept of sustainability, itself the core conceptual basis of green management, because by that time it became both clear and urgent that resources were ephemerous and their availability in nature, for human usage, was constrained, due to their inherent limited stock. Consequently, human beings could not make use at exhaustion points to satisfy their current needs if they would still be considering such resources were required for future generations—following this obvious mismatch in both ambitions, debates have emerged aimed at finding the best theoretical and managerial approaches so that both ambitions could be calibrated (Glück, 2001).

Anderson (2004) highlights, as the rooting period (even though silent) for the environmental awake, the Industrial Revolution era originated in England in the late 1700s, once during this time the most devastating environmental harm in history has been originated. Hobsbawm and Cumming (1975) corroborate this perspective, stressing the relevance of this period in history as it crystalizes the milestone for the move toward the capitalistic economy, where wealth and profit candied as the valued goals of individuals and corporations.

These initial inputs developed over time, reaching a turning point around the mid-1960s, being speedily boosted over the upcoming decades, as during such time societal changes provoked weighty negative effects on the environment, which impacted society's awareness impelling it to start expecting businesses to take an active stand on assuming great part of the accountability for the planet's environmental problems (Buchholz, 1993). Hence, this period marks the cornerstone for sustainability debates' upsurge, as a result of the true environmental movement instigated by that time. All societal structures and agents, namely organizations and institutions, then found themselves impelled to accept the challenge of incorporating sustainable management initiatives in all of their functions, facing a precipitous learning curve when adopting such strategies, policies, and practices (Nattrass & Altomare, 1999).

A next era of environmental awareness traversed the 1980s, as identified by Nattrass and Altomare (1999), when tragic environmental events took place. Those problems crystalized as vibrant alerts demanding that all societal agents (in the first place organizations and institutions as a collective and organized people structure) should go beyond legal compliance to truly assume themselves as respectable corporate citizens (McGuire, 1963). Following this line, in this period we globally witnessed the growing environmental deterioration due to climate change, which harmed environmental and human welfare, as well as companies' financial records. Thus, organizations gathered efforts to implement management practices and policies toward sustainability as means to proactively anticipate and best manage such scenarios.

Strategic questions surrounding environmental issues gained remarked urgency over time. As a consequence, sustainable management turned into an imperative vector in all agents' daily actions. Thus, the 1990s establishes a refined era of environmental consciousness, in which the term "eco-efficiency" has been coined, as this marks the period in which organizations began to ascertain that sustaining the *status quo* would not lead to a fruitful future. In fact, this period not only consubstantiates a more effective and visible action, but also a deeper awareness by all agents in regard to the environmental cause, given the fact that such approach could help them in obtaining a competitive advantage over competitors (Porter & van der Linde, 1995).

Nattrass and Altomare (1999) suggest the new millennium as the final and current era of environmental awareness, by observing that it has deconstructed the idea that environmental and organizational performances are two completely separated and irreconcilable fields. Instead, organizations are increasingly acknowledging the positive outcomes, even if in regard to their financial performance, that can be originated if adopting sustainable management policies and practices in the daily life. In this era, by assuming environmental issues as a chief concern in all functions, organizations remain competitive in their markets and clearly stand out a message to be socially responsible and conduct accordingly to moral standards, thus alongside aligning business and environmental goals.

Based on the theoretical background above provided, Taylor (1992) defines sustainable management as a management concept, given the fact that its core is peremptorily demanding the commitment of all the organization's stakeholders, as they recognize being an active part in the community, consequently assuming their responsibility quota on environmental conservation, sustainable development, and societal welfare. Thus, the concept imprints an holistic approach, as it involves conceiving the organization in its entireness instead of a conjunction of fragmented smaller entities. This means, such approach entails, as a vital premise, the fully and straightforward acknowledgement that the agent (people, organization, and institution) is an integral part of the community in which it locates and develops. This way, and by an external spill-over logics, it also impact the broader societal context (Van Marrewijk, 2003). Subsequently, it is imperative to manage the organization based on the long term, implementing an all-inclusive approach.

Truth is environmentalism indeed forces organizations to expand their corporate time-horizons because it impels them to be ahead of the public opinion. Consequently, under a sustainable management paradigm organizations are enhanced to commit to clear and strict quality performance standards in the broad scope of the organization's activities, developing fruitful and close relationships with all stakeholders, emphasizing the customer experience as vital, and sustaining motivation and finding creative solutions through the constant engagement and commitment of/toward employees, by acknowledging employees are vital not only to the organization's profitable performance but also to successful corporate change (Fernandes & Machado, 2020). Indeed, under this management paradigm organizations seek to improve the quality of all activities under the organization's (and its subsidiaries) scope, in all ranges that can be affected, directly or indirectly, being it on the short or on the long run (Málovics, Csigéné, & Kraus, 2008).

In a communion with the inputs provided by the abovementioned authors, Nelson (1998) and Zairi (2000) suggested an approach based on three elements aimed at creating and developing added societal value, as such arguing that organizations that have taken active steps directed to a true headway into the adoption of a sustainable management paradigm tend to progressively and intensely reveal four characteristics, namely: (1) they rely on value-based transformational leadership; (2) they objectivate a commitment to constant learning and innovation via the establishment of global networks and partnerships; (3) they establish, deepen, and reconfigure stakeholder linkages, hence taking advantage of mutual benefits via various modes of relationships; and (4) they make use of an extensive assortment of financial and nonfinancial performance measures reinforced by auditing, verification, reporting, and recognition systems, and this aimed at leveraging their performance.

Based on the inputs above presented, we can advocate that sustainability roots on three pillars, namely economic growth, ecological balance, and social responsibility. Furthering this assumption, we can also summarize that sustainable management consists, in its very essence, of an incessant process of assessment and improvement through the construction of environmental and management excellence, which derives on the assumption that it requires a deeper and ambitious approach to organizational actions as, at its very core, it impels to changing attitudes, structures, policies, and processes (Taylor, 1992).

Herewith, we can defend that sustainable management provides a refined management approach to conceive the contemporary and intractably complex relationship between business and society, not understanding corporate success, environmental conservation, and social welfare as a zero-sum game, instead analyzing such correlated dynamics as a win-win-win relationship. So, this refined approach clearly stresses the evident social character of economics, actively presenting it as a social science, which had been sometimes overlooked or misunderstood in classical approaches. Accordingly, we can clearly see that this conceptual umbrella is determinedly grounded in the sought of a long-term perspective in regard to human impact on the environment, and expanding this line, on individual and collective human impact on societal global development, which is impelled by the premise of attributing equal relevance to financial and economic efficiency, human resources, people holistic well-being, and environmental conservation (Lépineux, Rosé, Bonanni, & Hudson, 2016), by understanding the relevance of building sustainable businesses as a means for economic growth.

The advancement of time generalized this concept, adding a plethora of ideas for its interpretation (Zink, 2005). Indeed, in the management field we assist to an increase of empirical research and theoretical literature on Green Marketing (Peattie, 1992), Green Accounting (Owen, 1992), Green Retailing (Lai, Cheng, & Tang, 2010), and Green Management topics (McDonagh & Prothero, 1997). However, the core of all conceptual discussions around sustainability remains firm and in the present times absolutely highlighted as an imperative issue: it presents a dyadic dynamics, as it refers to both human beings' dependence on the environment and their reaction to global changes in such environment via its present behavior in the economic, social, and ecological domains.

Following this line, academic debate and empirical research still call out for the need of a more specific and widespread conceptualization for sustainable management, even though researchers fully acknowledge that it is complicated to crystalize a consistent and comprehensive definition on the first instance because we are analyzing and trying to operationalize a relatively new term (Florida & Davison, 2001).

Benefiting from the background above explained, it then becomes clear that the demands toward sustainable development are continuous, so they are exerted not only for this generation, but also for the upcoming ones (Marcus & Kaiser, 2006). From this premise, we can derive the concept of sustainable development, which consists of the strategies, policies, and practices designed and implemented in order to ensure the capacity of meeting the needs of the present agents without compromising the ability of the future ones to satisfy their own. Hence, the foundation on this concept relies on the conceptual interlinking of the environmental, social, and economic spheres via a "triple bottom line" (Elkington, 1994), whose development should, then, be deeply intricated and harmonious.

Therefore, we can conceptualize green management as the strategy an organization adapts to conceive and act toward environmental management, aimed at best facing environmental challenges, acknowledging that organizations essentially need to balance the industrial/business growth with the environmental conservation, so that we can reach the twofold objective of economic growth and global societal development (Daily & Huang, 2001).

9.5 Toward a circular economy

According to Wilkinson et al. (2001), two issues are crucial when debating sustainability, and it is in their interconnection that the debate itself can be furthered in order to understand if social efficiency is possible to be achieved and enhanced alongside to economic efficiency, via a continuous spectrum and a correlated relationship, forming a symbioses, or if such argument arises as nothing more than utopia, alluding to Fernandes and Machado (2021). So, the first issue addresses the changes that organizations (as people ordered and structured conglomerations) need to undertake in order to become agents of sustainable development, providing effective inputs to the implementation of a development model that meets the needs of current generations without harming those of the future ones. The second one entails the pursuit of an holistic management model, which then encourages people development, helps to achieve the organization's sustainable objectives, and leaves a positive footprint on the development of the social (national, international, and global) subsystems in which it operates (Hart, 1997).

This exercise culminates on this premise: the development of a society, at a national and global scale, does not depend only on the production potential of its economy, this to say, on the conventional and blind interpretation of the output maximization goal. Hence, to this approach it is imperative to acknowledge that economic, financial, management, political, and social actions are morally guided (thus, also limited) by the framework of values intrinsic to the people who design, implement, and perform such actions, and this is influenced by the surrounding context (Bento, 2011).

So, this approach consists of a useful theoretical and managerial perspective in order to understand management and economics both as conceptual and practical approaches from a point of view that is not only technical, but also imbued with a moral and ethical imprint, via a biunivocal and dynamic relationship (Fernandes & Machado, 2020, 2021). It indeed entails that it is not only convenient, but ultimately urgent and correct, to deal with these spheres and the issues from there derived in a unified and synergic flow, and crucially not in a trade-off logic (Svendsen, 1998). Therefore, a cooperative (and not a conflicting) dynamic is created between self-interest and collective interest: although often economic, financial, and social actions are based on utilitarian values, the complex system created by the spheres' connection should always be understood as having a subsidiary role vis-à-vis higher moral imperatives (Bento, 2011), which encompass the interconnection between efficiency and well-being, in a logic of leveraging aggregate utility, i.e., oriented—as the ultimate goal—toward social cohesion.

This proposition is sustained on the argument of Oliveira and Holland (2017), advocating that a radical inversion of the economic productivity principle will be vital for the interconnection of the two concepts: social productivity must, then, be engrained on the inverse of economic productivity, achievable through transformational leadership.

In this respect, further bold topics arise within the scope of welfare economics or social economics, complexifying the concept and the very scope of action of politics, economics, and management. We fundament that assumption based on the argumentation of Bento (2011), who postulates that improvements in social welfare are provided when the intervention of a generation produces positive results via spillover effects not only to the present context, but foremost to the upcoming generations. For this reason, the multiplication of individual decisions consubstantiates a qualitatively different outcome than would be the one achieved by the mere sum of the individual results, this being so because the multiplication of individually rational decisions does not, indeed, guarantee that the aggregate result is rational. This, in the ultimate instance, may give *momentum* to the challenging and growing (not only theoretical but also managerial) conflict between efficiency (economic) and equity (social cohesion).

In a full communion to Fernandes and Machado (2020), we highlight that this argument is pertinent because it both derives from and furthers into a logic of complex interdependence, which, however, is currently confused and distorted in both conceptual and practical terms. This is because, at the present times, there is a tendency to blindly base the notion of social efficiency only on economic efficiency, the latter being based on a tenacious and unrelenting concern with market satisfaction and financial leverage vis-a-vis the overlooking people well-being. However, people are ultimately the agents of construction, development, recreation, and vitality of the economy, either at a national or at a global scale.

Reaching this point in the argumentation, the authors make a chief remark which sharply reflects the complexity of the discussion: Fernandes and Machado (2020) point out that the words *efficiency* and *economy* are often juxtaposed, but the words *efficiency*, *welfare*, *well-being,* and *society* are not so often juxtaposed; indeed, they are often presented as antithesis. In fact, it has been assumed that social welfare could only be achieved if an economy became, first, more efficient. Financial and economic competitiveness should precede social well-being and welfare. In other words, economic growth would be an absolute precondition for income redistribution. Nevertheless, and once again corroborating the argumentation of Fernandes and Machado (2020), we point out that this view is flawed, as it tends not to recognize (or at least not properly) the circularity intrinsic to expenditure and income. In this sense, we suggest that economic efficiency should proceed rather than precede social well-being and welfare, based on the idea that income redistribution sustains (rather than drains) economies.

Furthering the argumentation presented by the abovementioned authors, we defend that an efficient market is based on economic and financial criteria, competitive advantage, utility maximization, trade-off costs, and private gains. Therefore, the operations taking place in it will be aimed at satisfying consumer needs and preferences, so the focus will be on market innovation and (de)regulation to maximize such output. In turn, an efficient society is concerned with and sustained on social criteria, collaborative action, mutual advantage, and social overall gains. Operations will, then, be directed toward meeting social preferences and demands, namely education, health, safety, environmental preservation, and human rights, with a constant and firm aim toward the enhancement of better life patterns, for today's generations and also for the upcoming ones, via the promotion of quality of life. Therefore, the focus will ground on social innovation, so on the process and product/service innovations observed in the aforementioned sectors.

In this regard, four characteristics are assumed as vital standards: relationships based on trust; reciprocity and exchange; common rules, norms, and sanctions; and connections in networks and groups. Relationships based on trust lubricate cooperation and, consequently, reduce transaction costs between people, aspect which, by its end, requires common rules, norms, and sanctions warranting that all actors take an enlightened and equal role in activities, which will ultimately foster reciprocity among them, with positive effects on the connection and cohesion of networks (Coleman, 1988).

Therefore, we suggest that the interconnection of these two concepts—social efficiency and economic efficiency—can be woven through the notion of social capital, which captures the idea that social bonds and norms are fundamental for sustainability, presupposing an ideal of social cohesion, hence on such dynamics being implicit the concept of cooperation. Thus, we postulate that a circular economy is the appropriate theoretical and managerial framework to conceive such complex and intricate relationships, this to say, to understand todays' dynamics.

Once again alluding to Fernandes and Machado (2020), as a corollary of the argumentation so far exposed in this chapter we, then, posit that the very essence of a circular Economy as both theoretical and managerial approach implies a (re)orientation on the political and economic discourse.

Circular economy, despite being currently a popular concept promoted not only by several national governments and international cooperation platforms, but also by many businesses associations and companies around the world, seems to entail only a superficial and unorganized scientific theoretical and empirical content. Consequently, the concept has been presented (in both academic debates and managerial research) as a mere collection of vague and separate ideas from several fields and semiscientific concepts, thus lacking a deep critical analysis.

Korhonen, Honkasalo, and Seppälä (2018) critically consider the concept of circular economy from the perspective of sustainable development and its three dimensions (economic, environmental, and social), thus suggesting a critical and refined scientific approach to its business concept: in their words, circular economy consists, then, in an economy constructed from the social dyadic production-consumption systems, which maximizes the service obtained via the linear triad formed by nature-society-nature material and energy throughput flow. This is implemented and achieved through cyclical material flows, renewable energy sources, and cascading-type energy flows.

Not disregarding the fact that, unfortunately, there is currently no effective, harmonious, and solid international policy effort to integrate circular economy approaches, the fact is that the circular economy, in fact, operates on four levels (Geng, Sarkis, & Bleischwitz, 2019). Namely, products need to be designed in order to be recyclable and reusable, so they must be rooted on green supply chains and clean manufacturing methods. On the other hand, organizations (firms) need different and innovative business models, so that they can efficiently create value and establish partnerships or other cooperative platforms, foremost for those groups of companies and customers which exploit a particular resource. On a third vector, policies are urgently required in order to facilitate markets and the transaction of goods and services via fair and sustainable trade. Last but not the least, on a fourth point it becomes clear that sustainable materials, especially biomaterials, are key for economic growth and then, fuelled by this later, to boost overall equative societal global development.

So, from this comprehensive definition, circular economy triumphs as a promising concept, because it has been able to attract not only the national states and the international cooperation platforms, but also the business community to constant and consistent sustainable development work. To achieve sustainability, the global community must deal with refined problems which are assumed as transcultural, transdisciplinary, and transversal to present and upcoming generations (Marcus & Kaiser, 2006). So, and based on the chief premise which defends that community engagement is crucial, this way it turns clear that researchers need to collaborate more, so that the concept of circular economy evolves to be widespread and, hence, actively transforms from a conceptual proposition to an effective action effort. This way, it impels a joint debate and cooperative action between academics, politicians, industry, and nonprofit organizations, via a concerted discussion (Geng et al., 2019).

Innovative, comprehensive, and inclusive research approaches (Berkes et al., 2000; Common & Perrings, 1992; Costanza, 1987; Golley, 1994; Holling, 1994; Viederman, 1994) have, henceforth, emerged as a result, aimed at better understanding and overcoming the challenges as above described.

Based on the above definition, it is then possible to stress that a successful circular economy has powerful effects both to economic and social efficiency, given the fact that it contributes to all the three dimensions, as above mentioned, of sustainable development. Right away on the economic efficiency side, we can verify that the business application of the circular economy principles may provide fruitful inputs in helping organizations hedge against commodity price volatility, thus rebalancing the flows of goods and waste (Geng et al., 2019). Circular economy, consequently, limits the throughput flow to a level tolerated by nature and exploits ecosystem cycles into economic cycles by respecting their natural reproduction rates.

It should be stressed that the emphasis should not be placed exclusively on one of the vectors mentioned above. Highlighting the alerts already noted by Fernandes and Machado (2020), attention should be drifted to a strategy of constant interconnection between all of the vectors above exposed, weighing their relative relevance in light of a strategic reading of the context in. Circular economy is assumed, hence, as a holistic theoretical and managerial approach.

Nonetheless, Korhonen et al. (2018) also claim limits on the proposed conceptual definition of circular economy in light of environmental sustainability, stressing it as a dare call for future research impelling it to a direction that postulates that the concept itself needs to be able to solve such issues, so that this approach may effectively contribute to global net sustainability. These challenges concern thermodynamics, the definition of system boundaries, and further issues in the governance and management of the interorganizational and intersectoral material and energy flows.

Following this line, Geng et al. (2019) advocate that a global strategy for the circular economy must include the following points as a focus on the upcoming efforts.

Firstly, a global database should be developed in order to turn possible the capture of trends and links between resource uses; then, and based on this initial data set as a base working tool, a global platform should be established in order to effectively allow knowledge transfer, this way allowing the development of learning activities on the circular economy scope.

These initial efforts would then develop so that we would reach a point in which international alliances needed to be established in order to permit large-scale data collection, learning, and experimentation. These should be configured and structured in order to focus on governance and financial innovations that underpin transformations in pioneering cities and economic sectors, in these especially referring to water, energy, food, mobility, and construction (as they are core for social well-being and development, not only for present generations but also for the sustainability of the upcoming ones). Therefore, this line of reasoning also demands particular attention and effective financial help to the countries/regions that are assumed as traditionally reliant on exporting primary materials, this being designed and implemented so that they can evolve and transform to more efficient and, foremost, sustainable performance levels. This, then, requires that, in order to effectively impel the world toward a circular economy, accurate and sustainable standards for performance measurement, reporting, and accounting need to be designed, developed, harmonized, assessed, and implemented. Deriving from this, also concerted means of enforcement, dispute settlement and sanctions at an international scale should be considered, advanced, and applied.

This evolution, on the words of Geng et al. (2019), will then climax on the idea that, in the long run, we would be able to reach an international agreement on sustainable resource management, thus an effective, harmonious, and concerted step toward societal global development. This could, in fact, emerge as a result of joint globally coordinated bottom-up approaches that include, via an equal value function, all the stakeholders, hence simultaneously being originated and being able to support international reporting, decision-making on eligible extraction sites for natural resources, favorable trade patterns, and acceptable economic incentives.

9.5.1 People as the core asset for societal development and sustainability

From the above explained, we can clearly advance that people and safeguarding of their holistic well-being are a founding basis on the concept of green and sustainable management.

Today's management strategies, policies, and practices aimed at actively implementing sustainable organizations require an holistic and long-term focus for planning and action, which can indeed go beyond the conventional blind pursuit of financial performance (Wilkinson et al., 2001). This is in light of the awareness of the resources' scarcity itself, and in here, we stress not only the natural ones, but on the first instance people as a core asset as well. Hence, the argument arises stating that fostering organizational sustainability (specifically in the people subsystem) will become a survival strategy for organizations (Ehnert, 2009).

This is, therefore, a conceptual and managerial approach subject to a higher level of complexity, compared to that was observed in the paradigm of strategic management, because it emerged from and developed into a dynamic and turbulent context, in which organizations were threatened by the changing nature of the constant, accelerated, complex, and interconnected dynamics deriving from the globalization. Globalization indeed powerfully fuelled competition, cooperation, and interdependence between the kaleidoscope of societal fields of action, fostering a concern for value creation.

The core of this approach is, then, the deep and complex interrelationship between people management strategies, policies, and practices, and the financial performance of the organizations, since it was argued that people management strategies would provide profit to the business through people actions (Ulrich, 1997). Nevertheless, the issue in here arises from the fact that people management must be conceived as an open system, in the sense that such linear approaches are dangerous (parsimonious) because in here we are talking about a broader and vividly complex scope of action.

Therefore, the green management perspective we have been postulating along this chapter entails a management perspective in which the results achieved can be measured by an holistic assessment and a wide-ranged evaluation entailing, in an equalitarian value logics, the organizational, social, individual, and ecological levels, as this approach considers them to be profoundly articulated. Central to all these issues is the clear focus on the purpose of management practices, imprinting on them an indelible moral character, which was not the case in the strategic management paradigm, in vigor by then.

Thus, and alluding to Kramar (2014, p. 1084), this management paradigm may be defined by its vital vector on people development and safeguard, being therefore understood as the pattern of planned or evolving strategies, policies, and practices aimed at enabling the achievement of financial, social, and ecological goals, while simultaneously reproducing the people base over a long-term perspective.

This argument converges on one point: people as the model's core (and in here we can also derive it to focus the analyses on the HR function). So, it conveys that people can play a central role in building a more ethical organizational basis, this way able to fully integrate social aspects into its more traditional practices. Because it is based on mutual advantage, the usefulness of this perspective is based on the fact that it has relevant implications for the containment of social inefficiency derived from excessive pressure for economic efficiency (Holland, 2000). Furthering this line, we can advocate that such management approach entails the minimization of the negative impacts on the natural environment, on people, and on communities, thus acknowledging the critical enabling role of all societal agents in order to achieve such goals. Expanding this line of reasoning, we can then support that, as a vital vector in this approach, the organizational culture will be guided by mutual recognition and understanding, respect for plurality, truth, integrity, and a cross-cultural point of view. Consequently, this management paradigm will value sometimes conflicting perspectives and a cooperative leadership style, encouraging open and honest communication in direct flows, integrity, and implementing participatory decision-making and problem-solving processes (Weaver & Trevino, 2001).

On this line, we can postulate that the relational models of management may be crucial as background for this approach because they enhance the involvement of people in the decision-making process in society (for example, workers in the organization), obtaining their commitment and thus favoring the achievement of high performance levels through concern with the well-being of all stakeholders. Therefore, together, the factors that support and exhaust resource capital and institutional capital imply optimal organizational structural configurations, which include decentralized structures, incentive systems that reward innovations, operational levels architected into multidisciplinary teams to facilitate learning, formal evaluation systems, horizontal flows of technology and information, and the development of programs that emphasize people experience, skills, and knowledge (Oliver, 1997).

Agreeing with this argumentation, Ehnert and Harry (2012) posit that the sustainability debate at the corporate, institutional, and people levels pacts with practices and strategies which produce significant impact on an organization's natural and social resources and environments, and this, expanding the action scope, is able to actively influence not only the institution's and the organization's future management conditions and business environment, but also (in a more critical way) people and the potential to meet the needs of the upcoming generations.

The awareness of people as a chief asset is critical in consubstantiating the paradigm shift that roots on both the theoretical and managerial refined approach of green management, presenting it indeed as a reformist approach, challenging the conventional assumption that the purpose of management is the pure achievement of business outcomes. Rather, by incorporating the notion of negative externalities (Mariappanadar, 2012), it advocates, as core premise, that human capital development will act as an essential outcome of management strategies, policies, and processes, taking an explicit and firm moral stance about the desired outcomes in organizational practices.

Acknowledging the fact that, in order to act as a potential source of sustained competitive advantage, a resource must be valuable, rare, inimitable, not substitutable (Barney, 1991), relevant (Grant, 1998), and dynamic (Johnson, Scholes, & Whittington, 2009), we can then postulate that this model presupposes that the sustainable competitive advantage derives from the symbiosis between resource capital and institutional capital, this way effectively unwinding into positive impacts toward societal welfare and well-being, impelling the global development.

It is, then, clear that the main concern involves the long-term survival of the organization, as a means that then, via external spillovers, overall well-being may be achieved in order to positively affect the potential for worldwide development boosted by economic growth, and with this, we can have an opportunity to correct the criticism of "myopia" addressed to the conventional economic perspectives, which inflated the short-term business goals.

This renewed conceptualization derives, according to Kramar (2014), from the fact that this approach recognizes and values the influence of the institutional context(s) on the development of management strategies, policies, and practices. Indeed, while resource-based theorists assume that managers undertake rational choices bounded by uncertainty, information asymmetry, and heuristic biases, institutional theorists assume that managers generally elaborate nonrational choices bounded by social judgment, historical limitations, and the inertial force of habit. This is the background inspiration for all the argumentation we have so far constructed, throughout this article. Being conceived and managed under this theoretical paradigm, the organization will then be able to balance economic and normative rationality, since this refined conceptualization of the source of competitive advantage enables the interconnection of resource-based approaches with those others of an institutionalist nature.

Then, organizational outcomes (stressing the relevance of human and social ones) will be more comprehensive and emphasized vis-à-vis purely financial ones. Following this line, we then suggest the need for a greater degree of complexity and rigor when looking at this debate in the current economic, political, and social context, based on the assumption that it is worth noting that the value of people contribution to society depends not only on the organization or industry where they are inserted and add economic value, but also on certain factors, such as political and cultural, with different configurations worldwide. Only then will the organization be able to capitalize on "isolating mechanisms," aspects of resources that prevent other organizations from being able to obtain and replicate them (Mahoney & Pandian, 1992).

Thus, and not denying the value of the theoretical approaches developed so far, we suggest the relevance of adopting a more holistic perspective on this point, because we consider that a single source of competitive advantage may not always be sufficient to achieve enhanced organizational performance—several sources should be interrelated given the fact that, and in line with Mueller (1996), managerial practices per se cannot transform people knowledge and skills into a valuable strategic asset, because competitive advantage (and its sustainability) can only be achieved through the combination of people and its management, the latter irremediably integrated in a holistic conception of the environment where organizations operate.

Reaching this point on the argumentation, it is then possible to systematize that the sustainable competitive advantage of an organization will depend on its ability to manage the institutional context which frames its resource decisions, which, by its end, includes the internal culture of the organization, and wider influences not only of the national state but also of a broader global context, undoubtedly not forgetting as well the identity relations that define a socially acceptable economy. Firstly to clarify the terms, resource capital comprises the firm's valuable assets and skills; institutional capital entails the firm's ability to support assets and competencies that advance its value proposition, thus embracing the context surrounding the resources and strategies that enhance or inhibit the optimal use of valued resource capital. This way, and according to Oliver (1997), we suggest that a joint combination of those two different types of capital are vital to the creation and sustainability of the organization's competitive advantage in the current and future economy, thinking in a logics of safeguarding the well-being, framed on a green management paradigm.

As a corollary, it is then possible to suggest that, rooting the source of competitive advantage of an organization on the relevant, inimitable and nontransferable capabilities of its people (efficiently managed through strategies, processes, and practices by a conscious management team, and also fully embedded in the economic, political and social context in which the activities are inserted), organizations will be able to capitalize on the environmental idiosyncrasies and, this way, attain additional returns. These outputs will be possible because organizations will be able to balance economic rationality—motivated by efficiency and profitability—and normative rationality—choices induced by historical precedents and social justification (Kazlauskaitė & Bučiūnienė, 2008).

Synthesizing, the relevance of this perspective grounds on the fact that it embraces, in a synergetic logic, spheres which have previously been perceived as conflicting and even irreconcilable: material (physical resources and financial performance) and immaterial (people performance, and social performance). This means this model of analysis acts as a catalyst for change in academical research and political debate, as it redefines previous theoretical and managerial approaches, namely those that sustained the source of organizational competitive advantage on technological superiority, on marketing or advertising strategies and operations, or even on financial performance (revenue potential).

Nonetheless, it is worthy to notice that this approach, although progressively developing especially as of the last decade, does not yet have an extended and cohesive body of literature (Bansal & Roth, 2000; Pfeffer, 2010). In fact, even if it refers to an implicit pragmatic consensus, it is still arguable that most organizations are reluctant in creating this efficiency (Ehnert & Harry, 2012).

9.5.2 Connecting ecological economics, green management, sustainable development and circular economy: Corporate social responsibility as the synthetic vector

As per the above mentioned around ecological economics and green management, the current worldwide scenario (conceived as the extended context framing the circular economy postulates) is posing higher public expectations on organizations' moral standards, so that ethics is nowadays decisively present among the core criteria in assessing corporate performance. Thus, new rules of corporate conduct are increasingly being considered: legitimacy, governance, equity, public/private-sector relationships, employment, and environmental impact.

Organizations are—regardless of their structure, core business, dimension, and geographical scope—progressively moving toward the adoption of advanced environmental practices that underpin their environmental performance along with their organizational competitiveness, simultaneously being inspired and pressured by a growing trend stressing that, being organizations the main cause of environmental problems, they shall as well assume a key role in addressing those issues and effectively acting toward them (Fernandes & Machado, 2020). In regard to this, we can highpoint that CSR strategies and practices as a constituent of the new societal governance.

Numerous conceptual inputs have been established in order to conceptualize CSR, mirroring the influence of several theoretical approaches, namely agency theory, the resource-based view of the firm, institutional theory, stakeholder theory, and stewardship theory (for a review, see Aguilera, Rupp, Williams, & Ganapathi, 2007; Carroll, 1979; Matten & Moon, 2008; McWilliams, Siegel, & Wright, 2006; McWilliams, Van Fleet, & Cory, 2002; Moir, 2001; Orlitzky, Siegel, & Waldman, 2011; Van Marrewijk, 2003; Wartick & Cochran, 1985; Windsor, 2006). This way, defining CSR is, indeed, a complex task (Crane, McWilliams, Matten, Moon, & Siegel, 2008; Lockett, Moon, & Visser, 2006).

It is obvious that the theoretical and managerial approaches in terms of CSR greatly benefited from the abovementioned discussions around ecological economics, green management, sustainable development, and circular economy, finding its historical roots when the debates around those concerns also started. Hereby exposing a very synthetized summary, in the 1950s it is possible to tackle the first solid inputs in regard to the evolution of the CSR concept. Even though scarcer innovative classifications flourished during the 1980s, it was indeed throughout that period that we observed a growing and more complex empirical research and theoretical debates on the subject. In fact, and as stated by Eberstadt (1973), it was prominently during this time that organizations (specifically business corporations) started to be theorized as institutions at which a myriad of vast social obligations are recurrently addressed, and as a consequence, they are increasingly called out to play an active role to fulfill them, alongside the government. Hence, this period sets the cornerstone in which alternative topics—including corporate social performance, stakeholder theory, and business ethics theory—begin to settle in academical work, in the political agendas and in the civil society (Carroll, 1999).

Analyzing the concept and its manifestations at a macrolevel, it is imperative to acknowledge that the assumption of CSR fluctuates among countries, as it is contextualized by national institutional frameworks (Chappie & Moon, 2005). Nonetheless, the nucleus of this concept mirrors social imperatives, and it respects the social consequences of business success. Thus, it plainly addresses the business' moral purpose, as it entails clear responsibilities to help society accomplishing basic goals, however, not disregarding that business consists of and must remain essentially an economic institution (Steiner, 1971). Tuzzolino and Armandi (1981) advocated physiological, safety, affiliative, esteem, and self-actualization as organizational needs, suggesting, this way, a broader conceptualization of CSR by developing the need-hierarchy framework inspired on Carroll's (1979) early contributions, and presenting a reinterpretation of Maslow's (1982) hierarchy of needs.

Thus, and as reinforced by Matten and Moon (2008), CSR entails the coherent and expressed organizational initiatives embracing the business accountability as a means toward the proactive action aimed at the extensive societal good, subsequently embodying, as by an equal value logics, concerns with corporate social performance, stakeholder relationships, connections affecting financial performance, corporate citizenship, and new applications of business ethics (Garriga & Melé, 2004).

So, the argumentation so far exposed sheds light on the fact that organizations indeed dispose a discretionary power in regard to the decision of which CSR approach to design and execute. This to say, the adoption of CSR strategies is mainly rooted on the organizations' self-interest. Dey and Sircar (2012) conclude by considering that CSR efforts must be internalized throughout all the organization's structures, strategies, policies, practices, and agents, so, all along its value chain, aim that must be incited indeed as a condition to develop it as a sustainable business. In fact, setting the accurate strategy is crucial in creating synergetic value (Kurucz, Colbert, & Wheeler, 2008), as the overall resulting benefits—namely costs and risk reduction, profit maximization, increased market share, enhanced employee motivation, higher customer confidence and loyalty, sustained competitive advantage, and improved corporate reputation and legitimacy—depend at a greater scale on the complexity and commitment level toward such initiatives.

It then becomes clear that, in order to achieve such goals (being them organizational, business, political, environmental, and social), organizations recognize the prerequisite to involve all stakeholders in CSR policies and actions (Zink, 2005). Expanding this logics, and alluding to Smith (2003), we can present two distinct motivators for implementing CSR practices in a given organization—the *normative* and the *business* vectors: they are not mutually exclusive, as an organization's incentive to engage in CSR activities might emerge as a particular symbiosis configuration between both vectors, in a myriad of joint possibilities all along a spectrum (Friedman, 1970).

Obviously, some CSR strategies and practices are prompted by the pressure of compliance toward a legal foundation: CSR discussion has, nevertheless, been facing a reorientation, as it has been gradually impelled to debate how such commitment should be made (by which means, via which initiatives, with which actors), instead of questioning why to compromise to CSR (Smith, 2003). It is, nevertheless, crucial to highlight that organizations adhere to CSR strategies and initiatives in a voluntary basis. Davis (1973, p. 313) offers useful insights on the relationship between these *mandatory* and *voluntary* vectors underlying an organizations' adherence to CSR initiatives, supporting that social responsibility is exerted as a proactive dynamics, once it is indeed the organization's obligation to gauge the effects of its decisions on the external social system, so that it can, via a win-win approach, achieve social benefits sideways to the traditional business economic gains. This allows to understand the vectors' interconnection as the core of the CSR concept.

In light of the above, reviewing literature we can expose an assortment of CSR strategies and practices. *Implicit* ones consist of a response to, or replication of, the organization's institutional environment; *explicit* ones condense the outcome of an organization's deliberate, voluntary, and strategic decision (Porter & Kramer, 2006).

The issue in here is that the *normative* motivator locates the organizational impetus toward social responsibility on the organization's pressure toward compliance, so, to its willingness to behave accordingly to moral standards; the *business* case, by their end, focuses plainly on the idea of rational self-interest, hence considering that focusing on social responsibility may propel concrete positive outcomes toward the firm's sustainable competitive advantage, leveraging the organization's economic performance. So, we can then contend that CSR consists of a nebulous idea, which, nevertheless, shall be integrated and objectively operationalized in a managerial context. As a summary and acknowledging these two motivators and its reconciliation potential, Carroll (1979, 1991) structured the concept by distinguishing its core economic, legal, ethical, and philanthropic tasks and responsibilities. Hence, we can defend the importance of CSR by sustaining that, via such joint approach, it would turn into an active part of the business, subsequently, increasing its potential to provide long-term value for both the organization and the society.

Impelling the analysis at a larger spectrum, we posit that CSR emerges as the synthetic vector which actively objectivizes the conceptual propositions of ecological economics, green management, sustainable development, and circular economy, as at its core there is, firmly, the notion of sustainability, once the establishment of a corporate social contract imperatively bases the CSR strategies and practices (Bowie, 1991; Davis, 1983). Frederick (1960, p. 60) early advanced this idea, stressing the relevance of CSR by clearly understanding its long-term orientation and arguing that "the economy's means of production should be employed in such a way that production and distribution should enhance total socio-economic welfare." So, assuming utility long-run maximization as the organization's major impetus and goal, the premise then derives to the idea that a socially responsible manager would be capable of articulating an equitative balance among all stakeholders' interests and needs, proving that business should serve a wider range of human values. Davis (1960, p. 73) climaxes such argument, stating that "the avoidance of social responsibility leads to gradual erosion of social power."

As a summary, we can then see that the concepts of ecological economics, green management, sustainable development, and circular economy often present some conceptual confusion and overlap, based on the argument that the conceptual proliferation so far available in literature has often been operationalizing such approaches as if synonymous, which then reflect into implications on organizational strategies, management approaches, academical research, political agenda definition, and civil society debates toward the sustainability imperative. The corollary of this corresponds to the idea that CSR represents the microeconomic dimension of the macroeconomic concept of sustainable management; subsequently, CSR actively links the concepts above exposed, even though it only consists of a precondition for effective and efficient sustainable management. We, thus, suggest that CSR is the synthetic concept linking the above conceptual approaches, which means it consists of the objective means to managerially structure and implement the theoretical approaches set forward by the above concepts. In other words, business sustainability is a consequence of the application of CSR strategies, premise that requires CSR strategies to be internalized into the overall business strategy, embedded across the organization horizontally and vertically, all along its value chain, in order to efficiently fulfill their very purpose (Frankental, 2001).

The argumentation so far exposed then allows us to defend that all the conceptual and managerial perspectives presented emerge as consecutive evolutions of the previous approaches' core premises, assumptions, and notions. This to say, ecological economics, green management, sustainable development, and circular economy conceptual structures and managerial approaches may be considered as "intrinsically linked," between them and also to CSR, as they all orientate toward the long-term viability of the economy, as a means to safeguard the overall sustained societal development, because at their core they prevail collective interests rather than the satisfaction of the mere conventional profit maximization, considering all economic agents are also social agents; thus, all parts involved on the economic process have an active footprint not only on the present world, but also on safeguarding the quality of life of the upcoming generations. Subsequently, people are clearly the core asset in all those approaches. This means CSR can be understood as the business contribution toward sustainable development. It is supported by a wide-ranging scale, both in what concerns to its geographical scope, as well as to what concerns the operational domains it impacts (economic, financial, nature conservation, education, welfare, wealth distribution, regional asymmetries, health care and safety, human rights, etc.).

9.5.3 The calling challenge: From welfare to well-being

The concepts of *welfare* and *well-being* are often considered together, thus presented and used in a juxtaposed way as if synonymous, especially *well-being* and *state welfare*. Nonetheless, and according to Taylor (2011), this debate can—in very simple and short terms—be set forward as regarding the conceptual (not only semantic, but also psychosocial) difference between "being well" (in which regards to well-being) and "doing well enough" (in which regards to welfare).

The production of the overall population's well-being and the maximization of such output consist of the core business of a welfare system, thus consisting simultaneously as both this approach's main assumption and goal (Dean, 2012). However, arguments can arise claiming that increased levels of social expenditure recorded by a given political system may ascertain and publicly expose a nation's inability to efficiently manage social problems, such as natural resources degradation and stocks' shortage, or high levels of unemployment. In fact, if an acceptable line of reasoning can be exposed demanding that welfare state policies and institutional architectural arrangements for human rights, education, health, safety, environmental conservation, and other domains around the overall social services may support in progressively enhancing the well-being for all citizens, another reasoned line of thinking can also be made to advocate that it is indeed not intrinsically required that a commitment to welfare unavoidably and automatically infers to well-being optimal output (Taylor, 2011).

Indeed, arguments about the size and role of the state, and, subsequently, of its intervention in the economic and financial system, have traditionally been a loud source of political tension in many market democracies, at least (and more frequently) in party and multiparty political systems. Thus, significant impacts are also addressed to the civil society in regard to this issue, as it both arises from and impacts on people, as societies contend to design and implement the suitable and acceptable (thus, legitimated) equilibrium between excessive and insufficient government involvement in people's everyday lives (Norman, 2010).

This discussion and its different nuances vary in accordance to the configuration of the political systems in cause, being that such configurations indeed differ in a kaleidoscope along a spectrum of openness to state intervention in the economy. Attempts to examine—and, foremost, to categorize the *welfare state* according to its idiosyncrasies and respective impacts on the overall affected subsystems—can be tracked back to the notable work provided by Titmuss (1958), and this being further developed by the noteworthy systematization provided by Esping-Andersen (1990), who was then able to present a typology of welfare regimes which mirrors the leading political movements and arrangements of political economy identified during the last century. These have been categorized and defined as being different configurations along the political spectrum, namely respecting to the Social-Democratic, Conservative and Liberal fractions of Welfare Capitalism.

Put into simple terms, the standard argument advocated in order to argue for smaller governments (by smaller, we refer to the governmental intervention in the public sphere) is addressed by the *right/center* of the spectrum, who claim that excessive state intervention on the economic domain may impose serious consequences on individual freedoms, thus sharply impact socially as it undermines the citizens' resilience and self-reliance. By expansion of this argument, it then becomes clear that all of these aspects might evolve into a negative effect on the population's well-being, thus being sharply scrutinized by the civil society (Welzel & Inglehart, 2010). On the other logics of argumentation, we verify that there is also a political discourse which stresses that people may feel discontented and concerned if ordinary and everyday life risks become internalized. In other words, understood as issues of personal failure and individual responsibility, instead of being addressed to the competent collective instances as being relevant and wide-scale social problems, through a concerted collective action (Taylor-Gooby, 2011). This last perspective focuses on the other extreme of the spectrum; thus, it is traditionally invested of a *left* character, advocating for the universal welfare statist argument.

Hence, in both cases we may see ground to address a possible deficit of trust (legitimacy) on the political government and its related institutions.

As set forward by Deeming and Hayes (2012), the contemporary challenges in regard to this issue, in the academical, political, economic, and civil debate, arise from the fact that we are facing the era of welfare retrenchment (Taylor-Gooby & Stoker, 2011). Subsequently, those issues are addressed as been not only complex, but also particularly opportune, being acknowledged the challenge currently facing social policy, as it tries to find a balance on designing the scope of maintaining established welfare state collectivism, despite cumulative liberal individualism (Karger, 2014).

As per the above paragraph, and not neglecting the connection of both concepts, this theoretical and managerial confusion is being increasingly put into test, via an interesting research trend which challenges to what extent does a welfare state regime classification indeed fundament and explain different configurations in well-being, when individual- and country-level characteristics are accessed and managed (Taylor, 2011). In fact, there is a substantial debate (both in the quantity of contributions and in terms of the complexity of its topics) in progress within theoretical and managerial conceptions and models around social policy in order not only to assess, but also to systematize and best operationalize the significant semantic and managerial differences between *welfare* and *well-being*, focusing, subsequently, on the sense in which *happiness* assumed as a key vector acting as the best measure—or, at least, as a sufficient measure—to track and evaluate the output and the progress of either (e.g., Dean, 2010; Dolan & Metcalfe, 2012). So, this new approach aims at reviewing the conventional argument in economic and political debates which claims that the aggregate level of well-being is expected to be higher within social democratic welfare states, which, subsequently, originates mechanisms which turn possible its distribution to be more equitable (e.g., Castles, 2010; Kvist & Greve, 2011; Wilkinson & Pickett, 2010).

The well-being of citizens is, thus, widely accepted as to be understood and measured in terms of the degree to which people live happy lives and foresee its continuity. The assumption rooting this argument states that, whether individual citizens are willing and feel they can afford the relative costs of their actions and assume accountable for them, they indeed see that the conditions are correctly set so that they can sustain the trade-off of scarifying some of their individual freedom (and the accomplishment of their private aspirations, needs, and demands), in order to preserve or even extend the collective welfare, hence acting aimed at promoting the social good. So, from here we can evidently detect the frequent and complex dilemma facing modern societies.

Therefore, and according to Deeming and Hayes (2012), discussion can be made in order to sustain that *happiness* might add an interesting input toward the enrichment of the social policy debate.

Even though, the truth is that, reviewing literature, it stands out that scarce studies, both focusing on theoretical discussions as well as on empirical research, have tried to explore and synthetize the complex relationship between well-being and the political intervention in the economies of the contemporary democracies. Most of the approaches derive from correlation analysis, postulating a link between the size of the welfare state—measured by its social security expenditures—and the level of well-being observed in its citizens. Thus, the size of the welfare state has been conventionally used in academical research as a proxy variable to assess the *welfare effort* and the degree of *state intervention* in the public sphere (Veenhoven, 2000). So, the contributes on this topic so far condensed in the literature mainly help to explain how the welfare of individuals in different types of a welfare state has fared over time, drawing on socio-demographic panel data (Goodin, Goodin, Headey, Muffels, & Dirven, 1999), on variations in health (Bambra et al., 2010; Eikemo, Bambra, Judge, & Ringdal, 2008), on the specific configurations on the social services (Alber, 1995; Anttonen & Sipilä, 1996; Engster & Stensöta, 2011), and on the relationship between welfare regime and public attitudes toward social welfare and redistribution (Andreß & Heien, 2001; Svallfors, 2011; Taylor-Gooby, 2011).

9.5.4 Societal overall well-being as the nuclear (and hybrid) input and output

Achieving an effective alignment between environmental/social management and organizational strategies, policies and practices requires a deep understanding of the alternative paths that sustainable organizations may take in the current world, on the first instance considering that their construction and constant development require a multidimensional vision that encourages superior organizational performance, which is, nevertheless and imperatively, subject (limited) to ethical standards—this to say, to societal moral scrutiny (Hoffman, 2007).

Hence, we can defend that the argumentation all along this chapter provides useful insights (being both inspired and inspiring) to the debate around social innovation.

According to Saul (2010), social innovation consists of a conceptual archetype which involves four key elements, namely: (1) designing and operationalizing an intentional business strategy; (2) leveraging core business; (3) creating new value (economic and social); and (4) enhancing positive social change. As by this twofold means and goals, social innovation hence changes the way in which organizations conceive and act toward social change, transforming simple philanthropy into an effective business opportunity, in the sense that it is via this approach (then objectivized into strategies, policies and practices) that organizations are indeed able to design and implement creative paths which originate, develop, and reconfigure profitable business opportunities from unmanageable social issues, such as education, health care, poverty, environmental conservation, human rights, gender equality, working rights, regional asymmetries, and global development.

This way, it turns clear that the conceptual and managerial approach we advocate via the argumentation exposed all along this chapter focuses on the societal overall well-being as—via an hybrid dynamics—its nuclear input and output. This is so because, via the operationalization of the CSR concept in organizations as core societal organized collective structures, the postulates of ecological economics, then evolved to the theoretical and managerial approach of green management, reached a complexity level which resulted into a sophisticated societal perspective embracing both social development and well-being, as well as economic growth and financial performance, which, consequently, crystalizes under the circular economy postulates. This new view, then, emerged as an harmonious fusion, originated by the evolution of the former approaches, thus creating a synthetic theoretical and managerial logics which we claim to be the most accurate one when thinking and acting in the current complex, diverse, accelerated, and interdependent world. As a root for this argumentation, there is the premise that the current economic crisis does not per se represent just another typical economic cycle, ground characteristic of the current capitalist financial and economic models: truth is, it represents a fundamental human recomposition, hence not only impacting on the economic, financial, managerial, environmental, and even political domain, but foremost impacting society emotionally, acknowledging people (future and upcoming generations) as the core asset toward worldwide development via an harmonious and sustainable path.

Therefore, this model intends to impel academics, managers, and politicians (all societal agents in an overall scope, being them on the economic, financial, environmental, political, and social domain) to envisage not only the contingencies they currently need to access and address, but also the challenges that may lie ahead, bringing into discussion the interconnectedness of internal and external inputs and outputs, as well as rising urgency on the analysis around the breadth and legitimacy of the served interests and its concealability. Thus, via the postulates of this approach, all societal agents are able to create an internal and external context that simultaneously focuses on and aims at sustainability. To this end, management of change will arise as a chief business, environmental, political, and social competence, which allows people to add value to their organization, natural environment, political system, family, peer group, etc. (to society, in overall terms), this way facilitating the development of core skills and knowledge in order to encourage the plasticity and the absorption of the constant volatilities which mark today's business, environmental, political, and social *milieu* (Ulrich, Brockbank, Yeung, & Lake, 1995).

To base this argument, we make use of Ellis' (2011) insights, who states that the contemporary business scenario configures as a complex network of dissimilar actors and interests, in a clear disarray, which then derive into the growing gap between rich and poor generations, as well as, intrinsically, in regional development asymmetries. In other words, in the 21st century, the rules are changing in regard to how to stay competitive in the business area, thus pressuring organizations, managers, and politicians to be more involved in dealing with social issues, so that they can indeed provide effective inputs toward overall societal well-being, this via concerted actions dynamized through an holistic approach.

Consequently, this concept emphasizes three dimensions of corporate performance: economic, social, and environmental (Steurer, Langer, Konrad,

& Martinuzzi, 2005). In fact, and as contended by Taylor (1992), even though the implementation of such approach requires up-front investment and often deep changes in the organizational structure and mindset, the fact is that the output provided crystalizes comprehensive advantages, both at the business and at the societal domain, worldwide. Indeed, this approach "rewards its adherents with: cost reductions and improved efficiencies; new marketing outlets; enhanced corporate image; opportunities to sell new products and services; an improved competitive position; a more dedicated and motivated workforce; and the ability to set the agenda for the industry and public policy" (Taylor, 1992, p. 674). So, the business advantages are real and measurable, aspect which may act as a relevant motivator toward the adoption of such approach.

This, henceforth, means that it is chief to notice that organizations are self-motivated to seek such purposes and to achieve the related output, and in this sense, we may argue that, given the fact that sustainable management provides outcomes which may be identifiable as public goods, organizations are, on their hand, not able to entirely appropriate such value (Teece, 2007). Consequently, we can contend that there are likely to be other reasons claiming organizations' attention to implement ecological, green, and sustainable management structures, strategies, practices, and policies (Marcus & Anderson, 2006), based on the argument set forward by Banerjee (2001) which instructs that the range of an organization's environmentally based approaches oscillates from reactive to proactive. In other words, the reality is that organizations can struggle, resist, or merely conform and comply with environmental principles, rules, and norms, or they can assess environmental issues proactively, embodying them at the core of their value chain as a chance to obtain, develop, and sustain a competitive advantage.

Not disregarding Banerjee's (2001) alert in regard to the motivations for ecological and sustainable approaches in organizations, so that they act framed into the postulates of a circular economy, we suggest that the sustainability agenda is not only fundamental to business strategy, but will, indeed, increasingly convert into a critical driver for worldwide homogenous economic growth and societal performance, as its core asset is people and all its efforts emerge from and direct toward the overall societal well-being. The author assumes so based on three key premises, namely: (1) economic development, (2) social innovation, and (3) sustainability. The first one claims that the main source of growth for multinational companies in the imminent times will be in emerging developing country markets, as proven that they favor and provide a positive contribution to economic development, regional asymmetries' blurring, and poverty reduction in these geographical areas. The second advocates that consumers' concerns about social justice, natural resources' conservation, human rights, poverty, and climate change are rising, impelling them to not only expect more from companies in terms of caring about these issues, but also actively and deeply scrutinizing their actions, as well as the ones derived from national and international governmental platforms, which, by their end, frame those others. The third one claims that all societal agents shall conceive the current world dynamics and all the related challenges as opportunities, rather than being hindered by the possible high costs associated. Hence, this vector of sustainability conquers *momentum* as globalization accelerates and the limits of the planet's resources are reached, so that academical debate and research, political discussion, and management strategies will be increasingly required to ensure the sustainability of their concepts and practices.

9.6 Conclusion

As a corollary, we suggest, based on Dryzek (2005) and Ferrão (2014), that current academical, political, and managerial discourses are indeed assumedly impregnated with challenging topics, acknowledging the global reconfiguration brought about by the complex and interdependent dynamics of globalization. Change and heterogeneity assume, with no doubt, as core topics transversal to all theoretical and managerial perspectives and approaches.

The argumentation exposed all along this chapter does not radically break with the quasicontinuity growth models in vigor so far. It acknowledges the conventional considerations derived from the "business-as-usual." But the truth is that it progressively distances itself from it and acquires specificity as it incorporates environmental concerns, centered on the search for economic growth marked by the more efficient use of resources, aiming at achieving a greater competitiveness of the economy. It assumes that it will only be possible when overall societal well-being is the prime goal of such approaches. This means our argumentation aims at fostering the discussion around economic assumptions at refined levels, reviewing them in order to evolve and to fully acknowledge, analyze, and manage the challenges addressed by the successive contextual reconfigurations, and being completely aware of the challenging threats brought about by the globalization and its impacts not only on the present generations, but also on the sustainability of the potential both for survival and for quality of life of the upcoming ones.

So, the argumentation on this chapter adopts the perspective of prosperity beyond economic growth, based on a criticism of the traditional basis of the capitalist market economy, assuming, therefore, as a basic vector the existence of clear limits to the exploitation of available resources. This way, the approach hereby defended acknowledges not only people but also nature as a core asset for economic growth and societal overall well-being, advocating the urgent and deep need for a profound review on the pursued purposes, due to a core assumption which defends the existence of clear limits to the exploitation of available resources (Ferrão, 2014).

So, the argumentation above described provides useful inputs to corroborate the "green pays" argument (Ambec & Lanoie, 2008; Crotty & Rodgers, 2012).

Therefore, and as a summary, we can see that, transversal in the literature review condensed in the argumentation above exposed, the concepts of ecological economics, green management, sustainable development, and circular economy often present some conceptual confusion and overlap, based on the argument that the conceptual proliferation so far available in literature has been often operationalizing such approaches as if synonymous, which then reflect into implications on organizational strategies, management approaches, academical research, political agenda definition, and civil society debates toward the sustainability imperative. The corollary of this corresponds to the idea that CSR represents the microeconomic dimension of the macroeconomic concept of sustainable management; subsequently, CSR actively links the concepts above exposed, even though it only consists of a precondition for effective and efficient sustainable management. We suggest that CSR is the synthetic concept linking the above conceptual approaches, which means it consists of the objective means to structure and implement the theoretical approaches set forward by the above concepts. In other words, business sustainability is a consequence of the application of CSR strategies, premise that requires them to be internalized into the overall business strategy and organizational value chain, in order to efficiently fulfill their very purpose (Frankental, 2001).

The argumentation so far exposed then allows us to defend that all the mentioned approaches emerge as consecutive evolutions of the core premises, assumptions, and notions present all along them. This to say, ecological economics, green management, sustainable development and circular economy conceptual structures and managerial models may be considered as "intrinsically linked," between them and also to CSR, as they all orientate toward the long-term viability of the economy, as a means to also safeguard the overall sustained societal development, because at their core they prevail collective interests rather than the satisfaction of the mere conventional profit maximization, considering all economic agents as social agents, thus assuming that all parts involved on the economic process have an active footprint not only on the present world, but also on safeguarding the quality of life of the upcoming generations. Subsequently, people are clearly the core asset in all those approaches. This means, as we said above, CSR can be understood as the business contribution toward sustainable development. It focuses on a wide-ranging scale, both in geographical scope and in what concerns the operational domains it impacts (namely, economic, financial, nature conservation, education, welfare, wealth distribution, regional asymmetries, health care and safety, human rights, etc.).

Therefore, all along this chapter we have argued toward a conceptual and managerial model which stimulates societal worldwide development via organizational sustainability, imperatively based on a refined conception of economics as having the environment and its resources as a core asset, conceiving the relationship between human beings and the natural environmental as one of complementarity.

We, thus, advocate that a circular economy (as embodying and operationalizing a synthesis between Ecological Economics and Green Management postulates) presents an alternative approach to current environmental and people management strategies, policies, and practices, as it identifies broader purposes and operational tools for economic, management, and societal development, by recognizing the complexities between the various actors and subsystems affected by organizational activities. This, undeniably, is assumed as an inclusive, comprehensive, and evolutionary approach, overcoming the traditional myopia which characterized the conceptual postulates and management models of mainstream neoclassical economics, because the argument we have put forward in no way means that organizational results (in key economic and financial terms) are no longer relevant: rather, we defend that their relative relevance needs (and, consequently, will) be revised, given the triumph of broader concerns at a worldwide scale, such as ecological footprinting, regional asymmetries, and social income redistribution, this way conceiving the current world as a complex and interdepend network of a kaleidoscope of actors and of interests.

It is clear that some green and sustainable organizational strategies, policies, and practices are already used in organizations, including work-life balance, forecasting future resource and capacity availability through workforce planning, and estimating the organizational ecological footprint.

Nevertheless, the insights condensed in this chapter alert that these measures need to be systematically adopted and to also be part of a broad and cohesive organizational strategy, marked by interdisciplinarity, informed by theories that make it possible to overcome ambiguity and grounded in rigorous measurement tools and feedback that links actions to results. This means CSR needs indeed to be assumed as more than just a trendy organizational concept, to embody a greater level of complexity and of commitment from the organization toward both the conventional organizational performance indicators (leveraging profits, leveraging return on investment, reducing operational costs, improving stocks' management, etc.), but also toward the overall global sustainable development, which means that the organization shall indeed not only develop efforts to attain their own aims, but also actively participate on the collective well-being, which means producing outputs which can indeed fall under the categorization of public goods.

This requires the awareness on the following assumption: designing the necessary transition strategies is different from mobilizing agents and decision-makers to trigger and manage the necessary change processes, which, by its end, is also different from effectively implementing and executing such efforts as a way to objectivize them into concrete actions which produce outputs. This is so because, as we progress along this spectrum of actions, we can plainly observe that more agents are involved, as well as more and increasingly delicate aspects are under analysis (e.g., the timeframe, the spatial scale, or even the constraint of successive management cycles).

Thus, via the postulates of this approach, all societal agents are able to create an internal and external context that simultaneously focuses on and aims at sustainability. To this end, management of change will arise as a chief business, environmental, political, and social competence, which allows people to add value to their organization, natural environment, political system, family, peer group, etc. (to society, in overall terms), this way facilitating the development of core skills and knowledge in order to encourage the plasticity and the absorption of the constant and relevant volatilities which decisively mark today's business, environmental, political, and social *milieu* (Ulrich et al., 1995).

Thus, exploring sustainable management and worldwide societal development as a strategic organizational topic points toward examining the means by which it affects the organization's competitiveness and profitability, and how it can be unified to the firm's strategic planning processes (Banerjee, 2002; Judge & Douglas, 1998): the argumentation in this chapter defends that the organizational sustainable competitive advantage roots on the organization's relationship with the natural environment, condensing a conceptual framework which encompasses the interconnected strategies of pollution prevention, footprint reduction, work-related legislation, healthcare and safety standards, product stewardship, and sustainable development.

With this chapter, we will be adding a fruitful contribution in responding to the calls in the literature aimed at integrating ecological economics, green management, circular economy, and sustainable development as a subject of research (Jabbour & Santos, 2008; Jackson et al., 2011), by means of not only plainly summarizing the interconnections and the dissimilarities between them, but foremost seeking to involve and fuse such approaches in order to update and refine them to indeed achieve the most accurate theoretical and managerial socio-political-economic approach to face the global current challenges, the ultimate conceptual framework to analyze the worldwide challenging dynamics brought about by the globalization.

By expansion of these contributions, we also defend that the insights condensed on this chapter as well offer a prolific contribution in order to expand the scope of strategic management toward sustainable management, so that it can plainly incorporate social issues, considering environmental conservation as a prime concern, thus aimed at also managing the public goods in order to leverage worldwide equitative economic growth as a means for the societal development, because people are the core of such approach (Osland & Osland, 2007; Wilkinson et al., 2001).

Therefore, the reported conceptual framework aims at informing, challenging, and enhancing reflection on how to develop economic models in a context that is indelibly marked by social concerns. This way, we present a twofold input—which are assumed both as a contribution derived from the argumentation all along this chapter and as a call out for future research on the area—because we state that multidisciplinary approaches are needed in order to understand the complexity and the heterogeneity of todays' world, given the fact that change and diversity are the current core vectors in the global scenario. Hence, the key implication of the proposed argumentation lies in the fact that an organization's ability to generate income from resources will depend primarily on its effectiveness in managing the context (internal and external). Therefore, the design and implementation of management models in this way will enable organizations to be seen as spaces for co-creating opportunities, in which each individual assumes vitally as a co-creator of solutions rather than a mere executor of tasks and functions (Fernandes & Machado, 2020, 2021).

Hence, and by expansion of this argument, the present chapter calls out the attention for the fact that it is indeed by developing awareness on all the involved stakeholders that it will be effectively possible to empower them in order to take an active role in society. This to say, and first of all, debates need to be made so that all agents fully understand the scope and the complexity of the dynamics taking place and the respective challenges in the current era, so that they can, then and via a joint action, design strategies which can reconcile all the interests involved, best managing them via a win-win game. So, an organizational culture that supports green management, this way framed under the nuclear basis of a circular economy, is one that encourages all the stakeholders to clearly stand out their points and to make suggestions that can ensure them as having the freedom to engage in activities to protect the environment and to foster societal well-being in a wide range of domains, both in an organizational level and in a personal level (Fernández, Junquera, & Ordiz, 2003).

This way, not only individual concerns will be attained, but foremost we will have a safeguard that overall social well-being will be achieved as the output of such strategies. Hence, it will be possible to produce collective goods alongside to private ones.

Therefore, we suggest the need for future research on sustainable competitive advantage, sustaining that it should focus on how resources are developed, managed, and diffused. And by expansion of this idea, we also suggest that, in line with Renwick et al. (2012), Green Human Resources Management assumed as both academical and managerial field with considerable potential as a research area. We can note that the path to the implementation of a Green Human Resources Management paradigm lies on the setup of supportive cultures for environmental management, and this is based on the fact that effective outcomes in terms of environmental management are achieved not just by making changes in production processes, products, or raw material, but rather by fully adapting the corporate culture so that we can have organizations profoundly embedded toward values which are effectively supportive of long-term sustainability (Kitazawa & Sarkis, 2000; Stone, 2000). Further empirical studies will, for example, be needed in order to understand how organizations can develop a team of managers proficient and fully aware of leading sustainability initiatives.

So, we highlight that efforts directed at identifying sources of both resource capital and institutional capital among competitors may shed additional light on the management of both, in order to foster the organizational sustainable competitive advantage, highlighting the relevance of longitudinal studies on the resources' development.

References

Aguilera, R.V., Rupp, D. E., Williams, C. A., & Ganapathi, J. (2007). Putting the S back in corporate social responsibility: A multilevel theory of social change in organizations. *Academy of Management Review*, *32*(3), 836–863.

Ajayi, S. I. (2001). Features-globalization and Africa—What Africa needs to do to benefit from globalization. *Finance and Development*, *38*(4), 6–9. English Edition.

Alber, J. (1995). A framework for the comparative study of social services. *Journal of European Social Policy*, *5*(2), 131–149.

Ambec, S., & Lanoie, P. (2008). Does it pay to be green? A systematic overview. *The Academy of Management Perspectives*, *22*, 45–62.

Anderson, R. (2004). Climbing mount sustainability. *Quality Progress*, *37*(2), 32–37.

Andreß, H. J., & Heien, T. (2001). Four worlds of welfare state attitudes? A comparison of Germany, Norway, and the United States. *European Sociological Review*, *17*(4), 337–356.

Anttonen, A., & Sipilä, J. (1996). European social care services: Is it possible to identify models? *Journal of European Social Policy*, *6*(2), 87–100.

Avant, D., Finnemore, M., & Sell, S. (2010). *Who governs the globe?*. Cambridge: Cambridge University Press.

Bambra, C., Gibson, M., Sowden, A., Wright, K., Whitehead, M., & Petticrew, M. (2010). Tackling the wider social determinants of health and health inequalities: Evidence from systematic reviews. *Journal of Epidemiology & Community Health*, *64*(4), 284–291.

Banerjee, S. B. (2001). Managerial perceptions of corporate environmentalism: Interpretations from industry and strategic implications for organizations. *Journal of Management Studies*, *38*(4), 489–513.

Banerjee, S. B. (2002). Corporate environmentalism: The construct and its measurement. *Journal of Business Research*, *55*(3), 177–191.

Bansal, P., & Roth, K. (2000). Why companies go green: A model of ecological responsiveness. *Academy of Management Journal*, *43*, 717–736.

Barnett, H. J., & Morse, C. (2013). *Scarcity and growth: The economics of natural resource availability*. New York: Routledge.

Barney, J. (1991). Firm resources and sustained competitive advantage. *Journal of Management*, *17*(1), 99–120.

Basu, K., & Palazzo, G. (2008). Corporate social responsibility: A process model of sensemaking. *Academy of Management Review*, *33*(1), 122–136.

Bento, V. (2011). *Economia, Moral e Política*. Lisboa: Relógio d'Água Editores.

Berkes, F., Colding, J., & Folke, C. (2000). Rediscovery of traditional ecological knowledge as adaptive management. *Ecological Applications*, *10*(5), 1251–1262.

Bowie, N. E. (1991). New directions in corporate social responsibility. *Business Horizons*, *34*(4), 56–66.

Buchholz, R. A. (1993). *Principles of environmental management: The greening of business*. New York: Prentice Hall.

Carroll, A. (1979). A three dimensional model of corporate performance. *Academy of Management Review*, *4*(4), 497–505.

Carroll, A. (1991). The pyramid of corporate social responsibility: Toward the moral management of organizational stakeholders. *Business Horizons*, *34*(4), 39–48.

Carroll, A. (1999). Corporate social responsibility—Evolution of a definitional construction. *Business and Society*, *38*(3), 268–295.

Castles, F. G. (2010). Black swans and elephants on the move: The impact of emergencies on the welfare state. *Journal of European Social Policy*, *20*(2), 91–101.

Chappie, W., & Moon, J. (2005). Corporate social responsibility in Asia: A seven country study of CSR website reporting. *Business and Society*, *44*, 415–441.

Cherian, J., & Jacob, J. (2012). A study of green HR practices and its effective implementation in the organization: A review. *International Journal of Business and Management*, 7(21), 25–33.
Chiang, C. (2010). Perceived organizational change in the hotel industry: An implication of change schema. *International Journal of Hospitality Management*, *29*(1), 157–167.
Christensen, P. P. (1989). Historical roots for ecological economics—Biophysical versus allocative approaches. *Ecological Economics*, *1*(1), 17–36.
Clark, C. W., & Munro, G. R. (1975). The economics of fishing and modern capital theory: A simplified approach. *Journal of Environmental Economics and Management*, *2*, 92–106.
Coleman, J. S. (1988). Social capital in the creation of human capital. *The American Journal of Sociology*, *94*, 95–120.
Common, M., & Perrings, C. (1992). Towards an ecological economics of sustainability. *Ecological Economics*, *6*(1), 7–34.
Costanza, R. (1987). Social traps and environmental policy. *BioScience*, *37*(6), 407–412.
Costanza, R. (1991). Ecological economics: A research agenda. *Structural Change and Economic Dynamics*, *2*(2), 335–357.
Costanza, R. (1996). Ecological economics: Reintegrating the study of humans and nature. *Ecological Applications*, *6*(4), 978–990.
Costanza, R., & Cornwell, L. (1992). The 4P approach to dealing with scientific uncertainty. *Environment: Science and Policy for Sustainable Development*, *34*(9), 12–42.
Costanza, R., & Daly, H. E. (1992). Natural capital and sustainable development. *Conservation Biology*, *6*(1), 37–46.
Crane, A., McWilliams, A., Matten, D., Moon, J., & Siegel, D. S. (Eds.). (2008). *The Oxford handbook of corporate social responsibility*. Oxford: Oxford University Press.
Crotty, J., & Rodgers, P. (2012). Sustainable development in the Russia Federation: The limits of greening within industrial firms. *Corporate Social Responsibility and Environmental Management*, *19*(3), 178–190.
Daily, B., & Huang, S. (2001). Achieving sustainability through attention to human resource factors in environmental management international. *Journal of Operations & Production Management*, *21*(12), 1539–1552.
Daly, H. E. (1985). Moving to a steady-state economy. In P. R. Ehrlich, & J. P. Holdren (Eds.), *The Cassandra conference: Resources and the human predicament* (pp. 271–287). Texas: Texas A&M Press.
Daly, H. E. (1994). Operationalizing sustainable development by investing in natural capital. In A. Jansson, M. Hammer, C. Folke, & R. Costanza (Eds.), *Investing in natural capital: The ecological economics approach to sustainability* (pp. 22–37). Washington, DC: Island Press.
Davis, K. (1960). Can business afford to ignore social responsibilities? *California Management Review*, *2*(Spring), 70–76.
Davis, K. (1973). The case for and against business assumption of social responsibilities. *Academy of Management Journal*, *16*(2), 312–322.
Davis, K. (1983). An expanded view of the social responsibility of business. In T. L. Beauchamp, & N. E. Bowie (Eds.), *Ethical theory and business* (2nd ed., pp. 94–97). New Jersey: Prentice-Hall.
Dean, H. (2010). *Understanding human need*. Bristol: Policy Press.
Dean, H. (2012). *Social policy* (2nd ed.). Cambridge: Polity Press.
Deeming, C., & Hayes, D. (2012). Worlds of welfare capitalism and wellbeing: A multilevel analysis. *Journal of Social Policy*, *41*(4), 811–829.
Deluiz, N. (2017). A globalização econômica e os desafios à formação profissional. *Boletim técnico do Senac*, *30*(3), 73–79.
Destek, M. A. (2020). Investigation on the role of economic, social, and political globalization on environment: Evidence from CEECs. *Environmental Science and Pollution Research*, *27*(27), 33601–33614.

Dey, M., & Sircar, S. (2012). Integrating corporate social responsibility initiatives with business strategy: A study of some Indian companies. *The IUP Journal of Corporate Governance, 11*(1), 36–51.

Dolan, P., & Metcalfe, R. (2012). Measuring subjective wellbeing: Recommendations on measures for use by national governments. *Journal of Social Policy, 41*(2), 409–427.

Drucker, P. F. (1993). The rise of the knowledge society. *The Wilson Quarterly, 17*(2), 52–72.

Dryzek, J. (2005). *The politics of the earth: Environmental discourses*. Oxford: Oxford University Press.

Eberstadt, N. N. (1973). What history tells us about corporate responsibilities. *Business and Society Review/Innovation*, (Autumn), 76–81.

Ehnert, I. (2009). Sustainability and human resource management: Reasoning and applications on corporate websites. *European Journal of International Management, 3*(4), 419–438.

Ehnert, I., & Harry, W. (2012). Recent developments and future prospects on sustainable human resource management: Introduction to the special issue. *Management Revue, 23*(3), 221–238.

Eikemo, T. A., Bambra, C., Judge, K., & Ringdal, K. (2008). Welfare state regimes and differences in self-perceived health in Europe: A multilevel analysis. *Social Science & Medicine, 66*(11), 2281–2295.

Elkington, J. (1994). Towards the sustainable corporation. *California Management Review, 36*(2), 90–100.

Elkington, J. (2006). Governance for sustainability. *Corporate Governance: An International Review, 14*(6), 522–529.

Ellis, E. (2011). The planet of no return: Human resilience on an artificial earth. *Breakthrough Journal, 2*, 39–44.

Engster, D., & Stensöta, H. O. (2011). Do family policy regimes matter for children's well-being? *Social Politics, 18*(1), 82–124.

Esping-Andersen, G. (1990). 4. The three political economies of the welfare state. *International Journal of Sociology, 20*(3), 92–123.

Fernandes, D., & Machado, C. F. (2020). How to look at organizations and human resource management in the economy of the future? In C. Machado, & J. P. Davim (Eds.), *Circular economy and engineering a new ecologically efficient model* (pp. 89–117). Switzerland: Springer.

Fernandes, D., & Machado, C. F. (2021). Stop camouflaging it in green: Do not confuse corporate social responsibility with sustainable management. In C. Machado, & J. P. Davim (Eds.), *Sustainable management for managers and engineers* (pp. 225–260). London: John Wiley & Sons.

Fernández, E., Junquera, B., & Ordiz, M. (2003). Organizational culture and human resources in the environmental issue: A review of the literature. *International Journal of Human Resource Management, 14*(4), 634–656.

Ferrão, J. (2014). (Coord.). *Que Economia Queremos?*. Lisboa: Fundação Francisco Manuel dos Santos.

Ferraro, F., Pfeffer, J., & Sutton, R. I. (2005). Economics language and assumptions: How theories can become self-fulfilling. *Academy of Management Review, 30*(1), 8–24.

Florida, R., & Davison, D. (2001). Gaining from green management: Environmental management systems inside and outside the factory. *California Management Review, 43*(3), 64–84.

Frankental, P. (2001). Corporate social responsibility—A PR invention? *Corporate Communications: An International Journal, 6*(1), 18–23.

Frederick, W. C. (1960). The growing concern over business responsibility. *California Management Review, 2*(4), 54–61.

Friedman, M. (1970). *The social responsibility of business is to increase its profits* (pp. 122–126). New York Times. September (13).

Gallagher, K. P. (2009). Economic globalization and the environment. *Annual Review of Environment and Resources*, *34*, 279–304.
Garriga, E., & Melé, D. (2004). Corporate social responsibility theories: Mapping the territory. *Journal of Business Ethics*, *53*(1), 51–71.
Geng, Y., Sarkis, J., & Bleischwitz, R. (2019). How to globalize the circular economy. *Nature Comment*, *565*, 153–155.
Glück, A. (2001). Das Prinzip Nachhaltigkeit–Zukunftsorientiertes Denken und Handeln in ausgewählten Lebensbereichen. *Politische Studien, Sonderheft*, *52*(1), 1–77.
Golley, F. B. (1994). Development of landscape ecology and its relation to environmental management. In M. E. Jensen, & P. S. Bourgeron (Eds.), *Vol. II. Ecosystem management: Principles and applications* (pp. 34–41). Portland, OR: US Department of Agriculture, Forest Service, Pacific Northwest Research Station.
Goodin, R. E., Goodin, R. E., Headey, B., Muffels, R., & Dirven, H. J. (1999). *The real worlds of welfare capitalism*. Cambridge: Cambridge University Press.
Gowdy, J., & Erickson, J. D. (2005). The approach of ecological economics. *Cambridge Journal of Economics*, *29*(2), 207–222.
Grant, W. E. (1998). Ecology and natural resource management: Reflections from a systems perspective. *Ecological Modelling*, *108*(1–3), 67–76.
Guthrie, J. P., & Datta, D. K. (2008). Dumb and dumber: The impact of downsizing on firm performance as moderated by industry conditions. *Organization Science*, *19*(1), 108–123.
Hammer, M., Jansson, A., & Jansson, B. O. (1993). Diversity change and sustainability: Implications for fisheries. *Ambio*, *22*(2–3), 97–105.
Hart, S. (1997). Beyond greening: Strategies for a sustainable world. *Harvard Business Review*, *75*(1), 66–77.
Hobsbawm, E. J., & Cumming, M. (1975). *Age of extremes: The short twentieth century, 1914–1991*. London: Abacus.
Hoffman, R. C. (2007). Corporate social responsibility in the 1920s: An institutional perspective. *Journal of Management History*, *13*(1), 55–73.
Holland, S. (2000). *Innovation agreements. Paper for the Portuguese Presidency of the European Council*. Lisbon: Office of the Prime Minister.
Holling, C. S. (1994). Simplifying the complex: The paradigms of ecological function and structure. *Futures*, *26*(6), 598–609.
Islam, A. (1999). Globalization and development revisited in the light of Asian experience. *Asia Pacific Development Journal*, *6*(2), 13–22.
Jabbour, C., & Santos, F. (2008). The central role of human resource management in the search for sustainable organizations. *The International Journal of Human Resource Management*, *19*(12), 2133–2154.
Jackson, S., Renwick, D., Jabbour, C., & Muller-Camen, M. (2011). State-of-the-art and future directions for green human resource management: Introduction to the special issue. *German Journal of Human Resource Management*, *25*(2), 99–116.
Johnson, G., Scholes, K., & Whittington, R. (2009). *Exploring corporate strategy: Text & cases*. Essex: Pearson Education.
Judge, W. Q., & Douglas, T. J. (1998). Performance implications of incorporating natural environmental issues into the strategic planning process: An empirical assessment. *Journal of Management Studies*, *35*(2), 241–262.
Karger, H. (2014). The bitter pill: Austerity, debt, and the attack on Europe's welfare states. *Journal of Sociology and Social Welfare*, *41*(2), 33–53.
Kazlauskaitė, R., & Bučiūnienė, I. (2008). The role of human resources and their management in the establishment of sustainable competitive advantage. *Inžinerinė ekonomika*, *5*(60), 78–84.
Khalili, N. R., & Duecker, S. (2013). Application of multi-criteria decision analysis in design of sustainable environmental management system framework. *Journal of Cleaner Production*, *47*, 188–198.

Kitazawa, S., & Sarkis, J. (2000). The relationship between ISO 14001and continuous source reduction programs. *International Journal of Operations & Production Management, 20*(2), 225–248.

Korhonen, J., Honkasalo, A., & Seppälä, J. (2018). Circular economy: The concept and its limitations. *Ecological Economics, 143*, 37–46.

Kramar, R. (2014). Beyond strategic human resource management: Is sustainable human resource management the next approach? *The International Journal of Human Resource Management, 25*(8), 1069–1089.

Kurucz, E. C., Colbert, B. A., & Wheeler, D. (2008). The business case for corporate social responsibility. In A. Crane, A. McWilliams, D. Matten, J. Moon, & D. Seigel (Eds.), *The Oxford handbook of corporate social responsibility* (pp. 83–112). Oxford: Oxford University Press.

Kvist, J., & Greve, B. (2011). Has the Nordic welfare model been transformed? *Social Policy and Administration, 45*(2), 146–160.

Lai, K., Cheng, T., & Tang, A. (2010). Green retailing: Factors for success. *California Management Review, 52*(2), 6–31.

Lee, K. N. (1993). *Compass and gyroscope: Integrating science and politics for the environment.* Washington, DC: Island Press.

Lenssen, G., Tyson, S., Pickard, S., Bevan, D., Castelló, I., & Lozano, J. (2009). From risk management to citizenship corporate social responsibility: Analysis of strategic drivers of change. *Corporate Governance: The International Journal of Business in Society, 9*(4), 373–385.

Lépineux, F., Rosé, J. J., Bonanni, C., & Hudson, S. (2016). *La RSE-La responsabilité sociale des entreprises-2e éd.: Théories et pratiques*. Paris: Dunod.

Lockett, A., Moon, J., & Visser, W. (2006). Corporate social responsibility in management research: Focus, nature, salience and sources of influence. *Journal of Management Studies, 43*(1), 115–136.

Lorenzi, N., & Riley, R. (2003). Organizational issues = change. *International Journal of Medical Informatics, 69*(2–3), 197–203.

Ludwig, D., Hilborn, R., & Walters, C. (1993). Uncertainty, resource exploitation, and conservation: Lessons from history. *Ecological Applications, 260*(2), 548–549.

Mahoney, J. T., & Pandian, J. R. (1992). The resource-based view within the conversation of strategic management. *Strategic Management Journal, 13*(5), 363–380.

Málovics, G., Csigéné, N. N., & Kraus, S. (2008). The role of corporate social responsibility in strong sustainability. *The Journal of Socio-Economics, 37*(3), 907–918.

Marcus, A. A., & Anderson, M. H. (2006). A general dynamic capability: Does it propagate business and social competencies in the retail food industry? *Journal of Management Studies, 43*(1), 19–46.

Marcus, A., & Kaiser, S. (2006). *Managing beyond compliance: The ethical and legal dimensions of corporate responsibility*. Ohio: North Coast Publishers.

Mariappanadar, S. (2012). The harm indicators of negative externality of efficiency focused organizational practices. *International Journal of Social Economics, 39*(3), 209–220.

Maslow, A. (1982). *Toward a psychology of being*. New York: Van Nostrand Reinhold.

Matten, D., & Moon, J. (2008). "Implicit" and "explicit" CSR: A conceptual framework for a comparative understanding of corporate social responsibility. *Academy of Management Review, 33*(2), 404–424.

McDonagh, P., & Prothero, A. (1997). *Green management: A reader*. London: Dryden Press.

McGuire, J. W. (1963). *Business and society*. New York: McGraw-Hill.

McWilliams, A., Siegel, D. S., & Wright, P. M. (2006). Corporate social responsibility: Strategic implications. *Journal of Management Studies, 43*(1), 1–18.

McWilliams, A., Van Fleet, D. D., & Cory, K. D. (2002). Raising rivals' costs through political strategy: An extension of resource-based theory. *Journal of Management Studies, 39*(5), 707–724.

Moir, L. (2001). What do we mean by corporate social responsibility? *Corporate Governance: The International Journal of Business in Society*, *1*(2), 16–22.

Mueller, F. (1996). Human resources as strategic assets: An evolutionary resource-based theory. *Journal of Management Studies*, *33*(6), 757–785.

Munda, G. (1997). Environmental economics, ecological economics, and the concept of sustainable development. *Environmental Values*, *6*(2), 213–233.

Nattrass, B., & Altomare, M. (1999). *The natural step for business*. Canada: New Society Publishers.

Nelson, J. (1998). Leadership companies in the twenty-first century: Creating shareholder value and societal value. *Visions of Ethical Business, Financial Times Management*, *1*(October), 21–26.

Norman, J. (2010). *The big society: The anatomy of the new politics*. Buckingham: The University of Buckingham Press.

Oliveira, T. C., & Holland, S. (2017). Economic and social efficiency: The case for inverting the principle of productivity in public services. In C. Machado, & J. P. Davim (Eds.), *Productivity and organizational management* (pp. 169–199). Berlin: De Gruyter.

Oliver, C. (1997). Sustainable competitive advantage: Combining institutional and resource-based views. *Strategic Management Journal*, *18*(9), 697–713.

Orlitzky, M., Siegel, D. S., & Waldman, D. A. (2011). Strategic corporate social responsibility and environmental sustainability. *Business & Society*, *50*(1), 6–27.

Osland, A., & Osland, J. (2007). Aracruz Celulose: Best practices icon but still at risk. *International Journal of Manpower*, *28*(5), 435–450.

Owen, D. (Ed.). (1992). *Green reporting: Accountancy and the challenge of the nineties*. London: Chapman Hall.

Peattie, K. (1992). *Green marketing*. London: Pitman.

Perrings, C. (1998). Resilience in the dynamics of economy-environment systems. *Environmental and Resource Economics*, *11*(3–4), 503–520.

Pfeffer, J. (1998). *The human equation: Building profits by putting people first*. Boston: Harvard Business Press.

Pfeffer, J. (2010). Building sustainable organizations: The human factor. *Academy of Management Perspectives*, *24*(1), 34–45.

Pfeffer, J., & DeVoe, S. E. (2012). The economic evaluation of time: Organizational causes and individual consequences. *Research in Organizational Behavior*, *32*, 47–62.

Porter, M. E., & Kramer, M. R. (2006). The link between competitive advantage and corporate social responsibility. *Harvard Business Review*, *84*(12), 78–92.

Porter, M. E., & van der Linde, C. (1995). Toward a new conception of the environment-competitiveness relationship. *Journal of Economic Perspectives*, *9*(4), 97–118.

Price, R. H. (2006). The transformation of work in America: New health vulnerabilities for American workers. In E. E. Lawler III,, & J. O'Toole (Eds.), *America at work: Choices and challenges* (pp. 23–35). New York: Palgrave Macmillan.

Rapport, D. J. (1993). Book review of man, nature, and technology. *Ecological Economics*, 7, 7–83.

Renwick, D., Redman, T., & Maguire, S. (2012). Green human resource management: A review and research agenda. *International Journal of Management Reviews*, *15*(1), 1–14.

Roome, N. (1992). Developing environmental management strategies. *Business Strategy and the Environment*, *1*(1), 11–24.

Sagoff, M. (1995). Carrying capacity and ecological economics. *Bioscience*, *45*(9), 610–620.

Saul, J. (2010). *Social innovation, Inc.: 5 strategies for driving business growth through social change*. San Francisco, CA: John Wiley & Sons.

Smith, N. C. (2003). Corporate social responsibility: Whether or how? *California Management Review*, *45*(4), 52–76.

Steiner, G. A. (1971). *Business and society*. New York: Random House.

Steurer, R., Langer, M. E., Konrad, A., & Martinuzzi, A. (2005). Corporations, stakeholders and sustainable development I: A theoretical exploration of business–society relations. *Journal of Business Ethics*, *61*(3), 263–281.
Stone, L. (2000). When case studies are not enough: The influence of corporate culture and employee attitudes on the success of cleaner production initiatives. *Journal of Cleaner Production*, *8*(5), 353–359.
Svallfors, S. (2011). A bedrock of support? Trends in welfare state attitudes in Sweden, 1981–2010. *Social Policy and Administration*, *45*(7), 806–825.
Svendsen, A. (1998). *The stakeholder strategy: Profiting from collaborative business relationships*. San Francisco: Berrett-Koehler Publishers.
Taylor, S. (1992). Green management: The next competitive weapon. *Futures*, *24*(7), 669–680.
Taylor, D. (2011). Wellbeing and welfare: A psychosocial analysis of being well and doing well enough. *Journal of Social Policy*, *40*(4), 777–794.
Taylor-Gooby, P. (2011). Security, equality and opportunity: Attitudes and the sustainability of social protection. *Journal of European Social Policy*, *21*(2), 150–163.
Taylor-Gooby, P., & Stoker, G. (2011). The coalition programme: A new vision for Britain or politics as usual? *The Political Quarterly*, *82*(1), 4–15.
Teece, D. J. (2007). Explicating dynamic capabilities: The nature and microfoundations of (sustainable) enterprise performance. *Strategic Management Journal*, *28*(13), 1319–1350.
Titmuss, R. (1958). The social division of welfare: Some reflections on the search for equity. In R. Titmuss (Ed.), *Essays on the welfare state* (pp. 34–55). London: George Allen and Unwin.
Tuzzolino, F., & Armandi, B. R. (1981). A need-hierarchy framework for assessing corporate social responsibility. *Academy of Management Review*, *6*(1), 21–28.
Ulrich, D. (1997). HR of the future: Conclusions and observations. *Human Resource Management*, *36*(1), 175–179.
Ulrich, D., Brockbank, W., Yeung, A. K., & Lake, D. G. (1995). Human resource competencies: An empirical assessment. *Human Resource Management*, *34*(4), 473–495.
van den Bergh, J. C. (2000). *Ecological economics: Themes, approaches, and differences with environmental economics*. Tinbergen Institute Discussion Paper, No. 00-080/3 Amsterdam: Tinbergen Institute.
Van Marrewijk, M. (2003). Concepts and definitions of CSR and corporate sustainability: Between agency and communion. *Journal of Business Ethics*, *44*(2), 95–105.
Veenhoven, R. (2000). The four qualities of life. *Journal of Happiness Studies*, *1*(1), 1–39.
Viederman, S. (1994). *The economics of sustainability: Challenges*. New York: Jessie Smith Noyes Foundation.
Vlados, C., Deniozos, N., & Chatzinikolaou, D. (2018). Global crisis, innovation and change management: Towards a new systemic perception of the current globalization restructuring. *International Business Research*, *11*(8), 9–29.
Vohs, K. D., Mead, N. L., & Goode, M. R. (2008). Merely activating the concept of money changes personal and interpersonal behavior. *Current Directions in Psychological Science*, *17*(3), 208–212.
Waddock, S. (2000). The multiple bottom lines of corporate citizenship: Social investing, reputation, and responsibility audits. *Business and Society Review*, *105*(3), 323–345.
Walton, S., Handfield, R., & Melnyk, S. (1998). The green supply chain: Integrating suppliers into environmental management processes. *International Journal of Purchasing and Materials Management*, *34*(1), 2–11.
Wartick, S. L., & Cochran, P. L. (1985). The evolution of the corporate social performance model. *Academy of Management Review*, *10*(4), 758–769.
Weaver, G. R., & Trevino, L. K. (2001). The role of human resources in ethics/compliance management: A fairness perspective. *Human Resource Management Review*, *11*(1–2), 113–134.

Welzel, C., & Inglehart, R. (2010). Agency, values, and well-being: A human development model. *Social Indicators Research*, *97*(1), 43–63.
Wilkinson, A., Hill, M., & Gollan, P. (2001). The sustainability debate. *International Journal of Operations & Production Management*, *21*(12), 1492–1502.
Wilkinson, R., & Pickett, K. (2010). The spirit level. Why equality is better for everyone. *Geography*, *95*(3), 149–153.
Windsor, D. (2006). Corporate social responsibility: Three key approaches. *Journal of Management Studies*, *43*(1), 93–114.
Zairi, M. (2000). Social responsibility and impact on society. *The TQM Magazine*, *12*(3), 172–178.
Zink, K. J. (2005). Stakeholder orientation and corporate social responsibility as a pre-condition for sustainability. *Total Quality Management and Business Excellence*, *16*(8–9), 1041–1052.

Index

Note: Page numbers followed by *f* indicate figures and *t* indicate tables.

FSC
www.fsc.org
MIX
Papier | Fördert
gute Waldnutzung
FSC® C083411

Kolibri 360 GmbH
Ferdinand-Jühlke-Straße 7
99095 Erfurt, Deutschland
produktsicherheit@kolibri360.de